To: My dear old
' The King
He ' Hon

CW01431068

UNDERSTOOD BACKWARDS

by

John Craven

The Memoir Club

First published in 2011 by
The Memoir Club
Arya House
Langley Park
Durham
DH7 9XE
Tel: 0191 373 5660
Email: memoirclub@msn.com

ISBN: 978-1-84104-524-5

Printed by Xpresslitho, Washington, Tyne & Wear

To my wife Barbara, our four children Sue, Pete, Dave and Rob and our eleven grandchildren Ben, Marc, Nicola, Mike, Caroline, Zoe, Jo, Charlie, Ashleigh, Harry and Jamie.

CONTENTS

List of Illustrations ix

Acknowledgements xiii

Foreword xv

Introduction xix

Chapter 1   The Thirties, Forties and Earlier 1

Chapter 2   The Fifties 12

Chapter 3   The Early Sixties 30

Chapter 4   The Rest of the Sixties 49

Chapter 5   The Seventies 62

Chapter 6   The Eighties 83

Chapter 7   The Early Nineties 94

Chapter 8   The Rest of the Nineties 104

Chapter 9   The New Millennium 117

Chapter 10   Retirement – and Second Childhood 126

Chapter 11   A Family Update and Some Reflections ... 140

List of Illustrations

Between pages 76-77

1. My father, Wilf, aged about 10 with his three brothers, Norman, Reg and Harold, 1913
2. Hilda and Wilf, 1928
3. 'The Chesters', Stanley. Wilf and Hilda's first home, 1930
4. Wilf and Hilda's wedding.
5. In siren suit with Sheena, 1939
6. Start of a lifelong addiction, 1939
7. With Mother, Father and Derek, 1940
8. Holmslyn, Whitley Bay, 1944
9. Holmslyn, 2011
10. Mowden Hall School, Stocksfield, Northumberland
11. Mowden Hall Rugby XV - aged 11
12. Durham School and Cathedral, 2011
13. Durham Regatta, record haul, 1954
14. Durham University VIII at Chester Head of the River, 1957
15. Durham University Rugby XV 1960/61- hirsute captain
16. University Agricultural Faculty at Newcastle, 1956
17. Professor Mac Cooper Dean of Agriculture, 1954-71
18. HTY 674 on tour in Spain, 1958
19. Pele Tower at Cockle Park, 1962
20. Dr Barbara Cooper, 1961
21. John Craven, an unlikely suitor, 1961

Between pages 108-109

22. Our wedding day, 1962
23. Figures and farming, 1965
24. Farm Management Consultant, Cheshire, 1966

25. Spy Hill Farm, 1965
26. Spy Hill Farm, 1975
27. With Sue and Pete above the farm, 1966
28. Contract rearing dairy heifers, 1966
29. Harvesting barley with a broken arm, 1966
30. Consulting Officers Conference, 1977
31. With John Frappell on a consultancy visit to India, 1978
32. Family Christmas at Spy Hill with the Thompsons, Morrises, Coopers and Cravens, 1973
33. Family camping holiday in Ireland, 1976
34. Mac and Hilary's Golden Wedding celebrations at Spy Hill Farm, 1987
35. My father and step-mother on their wedding day, 1965
36. Hilary and Mac with their three daughters, Diana, Barbara and Cynthia, 1987
37. Family holiday in the British Virgin Islands, 1999
38. Skiing in Zermatt
39. Presenting the Genus-sponsored Perpetual Trophy to Murray Stevenson
40. Retirement gift presented by John Beckett

Between pages 140-141
41. Spy Hill Farm looking east towards Delamere Forrest, Autumn 2010
42. Spy Hill Farm looking west towards the farm and buildings, Winter 2010
43. The lake, early Spring 2004
44. Excavating the lake, 1996
45. Ben Craven supervising operations
46. Members of the A and B Teams
47. A Ride to Remember. The start at Lands End, August 2001
48. Ted Charlesworth presenting part of the money raised to Ken Roache
49. Chester Golf Club - Captain's Day, 2004

50. With Barbara and Lady Captain, Sheila Watson, dressed up for the Annual Ball

51. Captain's Drive into office, 2004

52. The 50th anniversary of the first Annual Dinner of the Honourable Society of Newts inaugurated in 1956

53. Summer Newts Meeting at Hawkestone Park, 2004

54. The old barn and cowshed before conversion, 2008

55. Spy Hill Lodge, 2011

56. Family group, 2010

57. Joe Bright, Brian McAdam and John Willis, 2002

58. Rapid Transit, Scilly Isles 2002

59. Grandchildren, 2009

Acknowledgements

A number of people have been kind enough to give of their time to read draft material during the writing of this memoir. Two of these are former colleagues who helped remind me of various episodes during my time with Genus. One is Steve Amies, whose thoughtful and objective opinions I have valued highly for many years. The other is John Beckett who was the company's first Chairman and has himself contributed much to its success in recent years. John was also kind enough to write the foreword for this book. A former Head of English at Durham School, and a particular friend of my brother, Derek Baty has known me for over sixty years. He was rather press-ganged into reading the draft material and I appreciate the help, professional guidance and encouragement he has provided.

Jennifer Soutter works as an editor for the Memoir Club and she has been most helpful in recommending changes to the layout of the manuscript as well as improving my prose and punctuation. I also add my thanks to Lynn Davidson, the MD of the Memoir Club for her patience during the whole process.

Within the family, my son Robin and eldest grandson Ben have read parts of the book as it has emerged and provided valuable opinions on readability – we will see what the others make of it in due course.

Finally, Barbara, my wife, has encouraged me to get on with it and I think feels it is a fairly accurate, if at times, slightly self-indulgent account of my activities since she met up with me some fifty years ago.

Foreword

This autobiography will be of particular interest to many dairy farmers in England and Wales. Much of it concerns John Craven's career within the Milk Marketing Board, spanning over thirty years, including its eventual abolition in 1994. Whilst there is little argument that the statutory monopoly created in the 1930s to protect milk producers from exploitation by the buyers of their product was a sensible and necessary piece of legislation, there are differing opinions as to how the organisation managed its affairs.

As a dairy farmer and cheese producer, I was more critical than most. The MMB monopoly was the only source of milk and also the sole agent for our cheese. Milk quality was poor. We were not permitted to innovate and were restricted to Cheshire cheese only, while Dairy Crest, the MMB subsidiary company, took our market. Moreover towards the end of its tenure some of the individual members of the Board seemed more interested in their own personal futures, rather than those of their constituents.

But this memoir concerns the farm production end of the business, rather than the administrational and political machinations of the industry. The narrative is written with the aim of holding the non-farmers' attention. It follows John's long career from trainee farm management consultant to Chief Executive of Genus, the successor company which arose from the ashes of the MMB. It is written within the context of the times and provides a revealing insight to the problems which arose and the individuals with whom he had to deal.

I first met John in the late 1960s when he visited my farm in Shropshire to discuss dairy herd management. He was enthusiastic, forthright and practical – I liked that, even if I didn't agree with all his recommendations. I watched his career progress but it wasn't until thirty years later that he approached me with regard to the possibility of joining the first Genus Board.

The cattle-breeding side of the MMB operation had been languishing for years. Dairy farms and cow numbers were in steady decline. The core business of artificial insemination was diving and worst of all the genetic potential of UK bred dairy bull semen had fallen so far behind the international competition that home sales were on the verge of collapse. The whole business needed re-structuring and some very tough decisions had to be taken. When John took over the reins in the early nineties he had the unenviable task of persuading the MMB board to approve the slaughter of the entire stud of some 600 bulls and to start again from scratch. In addition he had to inform them that it was necessary to close down the majority of field centres and make hundreds of staff redundant. It was a traumatic time and he did well to get his way. Many farmers had little confidence that Genus would survive those first few years.

John retired in 1996 and under a new Chief Executive, Genus has diversified and grown into a world leader in cattle and pig breeding technology. John was a farmers' man and he would be the first to say that he didn't have the City experience to lead the company into acquisitions and takeovers. But had he not managed the transition from the MMB to an independent plc, Genus would have had a short lifespan.

This story recounts a period of dairy farm history which is unique. It describes the challenges and the personalities of some of the major players seen through the eyes of someone at the sharp end.

In his introduction the author sets out his reasons for writing this memoir. He points to the genetic link which he has with his children and grandchildren and hopes that the story might help them better understand their own make up and personalities. I feel sure he will not be disappointed. It is told with conviction, honesty and humour and it gives me pleasure to recommend the book as an enjoyable read.

John Beckett

Introduction

It seems rather pretentious to write a book about one's own life but I have my reasons.

First, I would have loved to have been able to read and understand more about my own father's life. His hopes and fears; his successes and failures; his joys and sadness. So often successive generations never have the opportunity to appreciate the characteristics of their antecedents within the context of the time in which they lived. A better understanding might help them interpret their own make-up. After all, our genetic inheritance is a fundamental base from which we develop our own personalities. This memoir may fill some gaps and maybe even provide some pointers to those who follow on.

Second, and probably the main reason is sheer indulgence. I have been lucky enough to enjoy a full and rewarding life, although at times it did not seem so. I and my family have been born at a time which avoided involvement in war. I have a happy marriage and loving family. I have experienced a challenging and exciting career working within the dairy industry. I've been able to participate in a range of sports at a competitive level and I have a wide circle of good friends. In short, I have been exceedingly fortunate – many millions of people wouldn't be able to say that.

Thirdly, I rather enjoy writing and various people, whose views I respect, have been kind enough to say that they find my style readable. In 2000 I wrote a biography of Mac Cooper, my late father-in-law and professor at Durham University. The principal reason was to record for the family, a remarkable life, before time obscured his memory. I was pleased and surprised by the response I received from friends, farmers, academics and his former colleagues, both in the UK and in New Zealand. He influenced many lives.

Finally, this opus is a means of self-analysis. The well known aphorism, 'life is lived forwards but understood backwards' is so true. The result probably doesn't provide any answers but the pleasure of filling the small hours with positive thought and reflection makes sleepless nights more bearable.

CHAPTER ONE

The Thirties, Forties and Earlier

WE TROOPED SILENTLY into the room. My father had said that he had something to tell me. I knew it was pretty serious. We sat down. He said, 'I'm sorry to have to tell you that Mummy has gone to heaven'. I promptly burst into tears and so did the others in the room. I was eight years old. It was my half-term holiday from boarding school and although I knew my mother had been ill and was in a nursing home, it was a terrible shock to learn that she had died.

It was February 1946 and the Second World War had ended only a few months earlier. There were four of us in the 'front room' of a small house in Consett, County Durham – my father, elder brother Derek and my mother's brother, Uncle Harry who, with his wife Linda, lived there. It was a small terrace house with an outside loo in the back yard. Downstairs there was a sitting room and kitchen where the family lived. The 'front room' which was hardly ever used, contained all the best china in a glass-fronted cabinet as well as the period furniture, much of the surface of which was covered with ageing sepia photographs of long dead relatives.

From this distance, it may seem a harsh way to break the news but I know they must have agonised for hours on how best to deal with this tragic event. There were no women in that front room – they, my two aunts and maternal grandmother were in the sitting room awaiting the emergence of the men, and I was encouraged to be brave and keep a stiff upper lip. There was no lack of love or caring, it was just the way it was. I remember also, in the evening, as a treat, we went to the local cinema to see *Tom Brown's Schooldays*, not an altogether welcome reminder of my return to school in a couple of days.

No doubt Lady Bracknell would have considered me, like Ernest, careless to have lost so many close relatives in my early years: my memories of my mother, to say the least, are extremely vague, and I have no recollection of either grandfather or paternal grandmother. My father was therefore the sole parent for my brother and I. We were the focus of his life.

The thirties was a desperate decade beginning with the consequences of the Wall Street crash in 1929 and the onset of the Great Depression which caused such misery and suffering throughout the world. It also helped establish Hitler as the rising power in Germany and following his appointment as Chancellor, set in motion the chain of events which was to lead to the outbreak of the Second World War in 1939. As the Allies were pushed out of Europe and one country after another threw in the towel, it must have seemed that life in Britain was about to change forever. These were dark days indeed and whilst their worst fears were never realised, my parents must have wondered how we would survive as a family. The war and its aftershock lasted throughout the forties. Disillusionment with the old politics and the birth of the Welfare State under a new Labour Administration ushered in fundamental change.

Antecedents

The Cravens were in the building trade. Thomas Cass Craven, my grandfather, born in the 1870s, began a small business at the beginning of the new century in Stanley, County Durham, initially carting sand and gravel for the local pit owners. He had started work for his father who had been a builder and monumental mason in North Yorkshire and allegedly that family had built a war memorial in Headingley, Leeds. For which war is not clear. There is also a mention of a road bridge in Ripon and even the Leeds Town Hall. The more romantic members of that generation subscribed to the view that earlier ancestors included the famous Earl of Craven, notable for his initiative in helping to put out the Great Fire of London in 1666. However, it has to be said that no convincing evidence is to hand to support this speculation. Furthermore, his was a Catholic family and the second son was expected to go into the Church. Had that been the case, then many of those who sought to support this notion may never have been born.

Thomas Cass's business began to expand before the First World War and he started to build houses for those employed in the local coalmines. One such development assumed the name of Quaking Houses, which no doubt referred to the subsidence which was forever a hallmark of this mining area, as the buildings overlay a network of underground tunnels. The local pit owners preferred to buy the surrounding land rather than be liable for future claims for compensation from unfortunate homeowners. But despite these difficulties, the building business was prospering and snapshots in the 1920s show staff outings and picnics for some twenty or more employees and their families.

Thomas Cass married Sarah Brewis in the late 1890s and they subsequently had four sons. Norman, the eldest, became a teacher, but died from TB at the early age of forty-five. The second son, Wilfrid, born in 1903 was my father.

He left school at thirteen and began to serve his apprenticeship as a bricklayer within the family firm. He was apt to remind us all in future years how he had to walk four miles to work in all weathers and be there by 6.00 a.m. His first job was to light the boiler and to make the tea for the rest of the staff when they arrived. Reg, the third son, also joined the business with a view to specialising as a joiner, whilst Harold, the youngest, continued his private education and eventually joined a local bank.

By the time the war clouds started to gather in 1913 the Cravens were moving up the social scale. They employed staff in the house, drove a motor car, and even went on holiday up to Scotland – a considerable adventure in those days. Thomas Cass was too old to enlist but as war became inevitable most of his employees left the firm to join the Durham Light Infantry, no doubt confident that they would be back within a few months having taught Jerry a lesson. The majority never returned. Those who did come back rejoined the firm, but the momentum had been lost. A new initiative was needed. This came in about 1920 when Thomas Cass acquired some land within the town of Stanley and proceeded to build a cinema – later called the Pavilion. The reason behind this departure from conventional house building remains unclear, as does the ownership of the theatre. However with the building trade woefully short of men, no doubt he was looking to diversify. Whatever the reason, it provided Wilf, as my father was known, with an entirely new challenge. He was to run the cinema as a separate business and rent the early silent films to show to the unsuspecting public. He had no experience but neither did anyone else as the film industry was just beginning. Shortly after this change of direction the building business folded and brother Reg joined Wilf at the Pavilion.

My father was clearly an energetic young man, one of four brothers who seemed, judging by the photographs in old family albums, to have little difficulty enjoying life. He was popular, had presence, a strong personality and a highly developed sense of humour. I don't recall his active participation in many sports other than cricket and golf, but as he grew older his passion turned to rough shooting and then fishing either for salmon or trout. He loved the countryside and in addition to camping expeditions, one of his main interests in the early twenties was motoring.

Firstly on belt-driven motorbikes and then a series of what now would be described as vintage cars. Although no formal training in matters mechanical, he seemed to have a natural ability and affinity for machines. Even in old age he was well known for his skill in fine-tuning car engines. To have the means to procure a motor car in those early days illustrates the relative prosperity which the Craven family must have enjoyed. There were frequent trips to the

Lake District and picnics in the countryside in all weathers. Furthermore to own one's own transport must have been an exceptional advantage when it came to impressing young ladies.

The film business

Films in the early twenties were simple love stories or classic cowboy adventures – black and white, silent and hopelessly naïve. But in those days they were tremendously exciting. People had seen nothing like them and long queues formed outside the cinema well before the doors opened. Although the films included subtitles so that the audience could follow the plot, the only sound was music. Each film had its own score and an orchestra would play the music in approximate time with the events on the screen. The more 'up-market' the cinema, the bigger the orchestra. There were plenty of musicians about as the Colliery Brass Bands were at the height of their popularity. To actually get paid for playing at the local cinema was a real bonus. Indeed music played an important part in the lives of the senior members of the Craven family.

Thomas Cass was an accomplished organist, having played in church since the age of fifteen and his wife, Sarah, also played the piano to a high standard. Not much of the talent rubbed off on their sons, however, although my father did attempt to learn the violin. Some would say that one of the prime motivations for my grandfather's investment in the cinema enterprise was that he could indulge his love of playing the organ. A handsome instrument was installed at the Pavilion and initially it formed an integral part of the full orchestra. In later years it became the sole source of musical accompaniment for the film and also provided the entertainment in the interval. Thomas Cass was in his element.

Cowboy westerns were probably the most popular films in the twenties, although one can imagine that a gunfight without hearing any shots might be rather unconvincing. So in attempt to enliven proceedings my father procured a small revolver which fired blanks. At the critical moment he would fire this at the back of the auditorium, often in quick succession. As might be imagined the first few nights this happened a number of the punters went into severe shock. But the word got round and the practice soon became the climax of the evening. On the rare occasions when the pistol shot coincided with the fall of the body, the audience would cheer enthusiastically.

The system for renting films in those early days involved the cinema proprietor agreeing a percentage of the money taken on the door. The better the film, the higher the percentage. It was a simple formula, but if the better film did not attract the crowds, then the owner was in for a lean week. Usually the film was shown for six nights (never on Sundays) and if the first night was a flop, the news spread rapidly round the town. This occurred on one particular Monday evening and the exit poll indicated that the thumbs were firmly pointed in a downward direction. My father acted quickly. He explained the impending disaster to the renter from whom he had hired this supposedly wonderful film and invited some ideas to rectify the situation. The renter was a man called Bill Travers, whose daughter Linden eventually became a Hollywood movie star. At the time Linden was beginning to make a modest impact in local repertory and her name was at least well known in Stanley. Bill's idea was that she would perform a prologue on the stage before the film actually commenced – this was sure to bring the crowds flocking in. The scene was to be a bedroom with Linden stretched languidly on the covers.

The dialogue required one other participant, who would be out of sight in the wings, but would conduct a conversation with the recumbent actress. The unseen part was to be none other than Wilf Craven and from that day forward he was anxious to remind anyone prepared to listen that he had appeared (more or less on stage) with Linden Travers well before she became a household name. In fact he became so infatuated with the future career of Miss Travers that he named his first son after her. My brother Derek's middle name was Linden.

As the twenties moved into the thirties and the Depression set in, the local cinema became a haven for many hundreds of people as their lifestyles began to suffer. Although money was exceedingly short, a night at the pictures became the highlight of the week. The films transported the audience away from the grim conditions they faced outside. Furthermore it was warm in the building and you could meet friends and talk about the films and the stars. It was a brief escape from reality. So it is little wonder that the company began to grow. Wilf and Reg opened similar establishments in a number of other small towns in the area, such places as Chopwell, Washington, Hetton le Hole and even Felling on the outskirts of Newcastle. By the beginning of the Second World War they had about fifteen of these operations and they were rapidly becoming well-known customers of the film-renting establishment. They had successfully diversified from building to entertainment and financial rewards soon followed.

My family

My father married Hilda Walton in 1930. She was the eldest daughter of Joe and Jessie Walton from Leadgate near Consett, a town within ten miles of Stanley. Joe was also in the building trade in the twenties, but his business became an early casualty of the Depression. In fact it was Jessie, originally from Scotland, who kept the family afloat. She was an accomplished seamstress and made clothes for sale to shops and private individuals. Their children, unlike the Cravens, all had secondary education, with Hilda attending Teacher Training College in Aberdeen. The newly-weds lived initially in Waverley Villas near Stanley but with the help of Harry, Hilda's younger brother, who had trained as an architect, and her father who still had plenty of contacts in the building trade, they built a bungalow on the outskirts of the town. Christened 'The Chesters', this was to be their home for the next fourteen years. The house still stands today but instead of being surrounded by fields as it was then, the area is now almost completely built up.

My brother Derek was born in 1931. His early memories were of a happy childhood, but often with a feeling of tension in the air. There seemed to be unpredictability in the relationship between his parents. Sometimes all was sweetness and light, other times there were heated arguments. It seemed to him, on reflection, that his mother was uncomfortable with the Craven side of the family, especially with her in-laws. She was undoubtedly intelligent and full of ideas for the expanding business. But this was a man's world and it was made plain to her that she was expected to concentrate on the home and family, not become involved in business matters. She was, however, not easily discouraged and established a café in the Pavilion, which sold confectionery, drinks and ice cream. It became a resounding success and the model was quickly replicated as the other cinemas came on stream.

She was also active in the town organising jumble sales and other fund-raising initiatives to help the multitudes of unemployed during the Recession. Derek recalled that his parents enjoyed a limited social life and it was only when with her own family in Consett that his mother seemed to relax and join in. Fortunately my father got on extremely well with Hilda's brother, Harry, and it was he who introduced him to shooting and subsequently fishing. Harry as a meat inspector knew a number of local farmers in north-west Durham and he engineered invitations to pursue rabbits, hares and the occasional pheasant on Saturday afternoons.

Not only did they enjoy the sport and the countryside but also in times of severe food shortage their efforts helped significantly towards supplementing the weekly meat ration. These excursions meant that much of the weekend was spent within the orbit of Hilda's family and only on relatively rare occasions did they visit Wilf's parents who had by now retired to Whitley Bay. Indeed Derek remembered spending a good deal of time on his own with Uncle Harry as he was dispatched to Consett on those occasions when his mother's health began to deteriorate.

I was born on 4 August 1937 at a nursing home in Newcastle. The story goes that my father took a telephone call whilst in Bainbridge's, one of the city's large department stores of that era, informing him of the new arrival. Derek, who was with him at the time, insisted that I should be returned forthwith as all had previously agreed that the fourth member of the family was to be a girl. Undaunted the two of them set out to look for a suitable present to mark the successful conclusion to the confinement and for some unfathomable reason decided on a giant sponge. But to their disappointment and surprise it was not received with any enthusiasm.

Early memories

My earliest memories concern the Second World War, which commenced two years after I was born. My father, who had unsuccessfully (owing to perforated eardrums) applied to join the Royal Navy had become a Special Constable and spent most nights patrolling the town of Stanley to ensure that the blackout was being strictly observed. As the war moved into the early forties the enemy bombing raids on the docks at Newcastle became a regular occurrence.

Our bungalow was about fifteen miles to the south-west of the city and in an elevated position providing a clear view of the exploding incendiary bombs as well as the dogfights in the night sky. A substantial air-raid shelter had been built at the bottom of the garden to which we repaired when the siren sounded. I remember vividly being carried down to the shelter in the middle of the starlit night, then lying awake listening to the distant explosions as the bombs began to fall. I also remember the occasional visits, on leave, of my Uncle Bill Brewis. He was my father's cousin and having emigrated to Australia had returned to join the Fleet Air Arm operating in support of the infamous North Atlantic Convoys to Murmansk. He looked resplendent in his RAF uniform and a true hero to us boys. Furthermore he seemed to take a real interest and played endless games of football on the lawn. But his greatest claim to fame was his snoring. In a two-bedroom bungalow this meant that no one, apart from Uncle Bill, got any sleep at all during his visits. Fortunately he survived the war and returned to Australia safely at its conclusion. We met briefly thirty years later when I was in Australia at a conference, although we had difficulty recognising each other.

We continued to live in Stanley as the war progressed and the tide began to turn in the Allies' favour. I attended the local primary school, of which I can remember very little. Pictures in old family albums point to a seemingly happy little boy, often dressed in a siren suit, similar in design but considerably smaller than the garment made popular by Winston Churchill.

In 1944 the family moved to Whitley Bay on the coast, about eight miles due east of Newcastle. This bracing town had taken a bit of a hammering during the war when German aircraft jettisoned their remaining bombs after raiding the Tyneside shipyards. My grandfather's generation would seem to have had a liking for the seaside. They regularly took their annual holidays at Scarborough about eighty miles to the south and although Whitley Bay lacked something of the Yorkshire resort's sophistication, it was, nonetheless, a big improvement on the landscape of north-west Durham, with its depressing coal mines and slag heaps.

Although my mother would have probably preferred to have moved nearer to her own family in Consett, she would also have been attracted to the idea of living at the coast, with its healthy sea air and cosmopolitan residents. They bought an attractive four-bedroom detached house in Holywell Avenue, reputedly designed by an architect related to the Bowes-Lyon family, for the then considerable sum of £2,500. Known as 'Holmslyn' the house had been built before the First World War and was to be my family home for the next sixteen years. Clearly the original owners had social aspirations, as there existed a call-bell system throughout the house in order to summon the servants to attend to their needs. The master display board in the kitchen indicated the room from whence the bell had been pushed, so it was with great glee that my brother and I ran from room to room pressing each bell in turn to both annoy and confuse anyone who cared to take any notice.

In fact quite soon after taking up residence my mother did employ a succession of so-called housemaids. It seemed this was more a mark of social standing than as a means of providing any significant help in the house. I remember the four of us sitting down to lunch in the dining room and Mother solemnly ringing the bell for the maid to come in and serve the next course of the meal that she had already prepared. None of them ever got it quite right and 'the incompetence of the maid' seemed to be a never-ending topic of conversation. These early days at the new house were happy enough but I do recall my mother being unwell on frequent occasions. My brother, six years my senior, remembered her extreme changes of mood, either elation or severe depression. To a small boy this was understandably confusing and it is with some sadness that I am unable to remember her displaying any real affection for either of us.

The schoolboy

My early schooling was extremely convenient. Holywell Avenue contained a number of small private establishments and I was duly dispatched to one of these, no more than a couple of hundred yards down the road. Unbeknown to my father the headmaster had plans for expansion and within a couple of years of our move he bought the property next door to Holmslyn. He procured the necessary planning permission to convert this large house into a school with new classrooms being built within a few feet of our boundary wall. The garden became a noisy playground and the whole development caused great distress and unhappiness to both my mother and father.

My father had had no formal education after the age of fourteen. Whilst he had done well to help develop the family business into an increasingly profitable enterprise he was determined that his children would have the opportunities which he had missed. In fact all the children born into the next generation had private education with the result that they mostly followed careers in the professions. Accordingly both my brother and I were shipped off to boarding school at the earliest opportunity. Derek went to a preparatory school near Hexham some fifteen miles west of Newcastle and later, at the age of thirteen, to Durham School. It was intended that I should follow him to Durham, but first a preparatory school needed to be identified and it was to Bow School, also in Durham, that I was sent in 1945 at the tender age of eight. It was not a success.

The headmistress, known as Fanny Lodge, was the archetypal battleaxe. I remember her as a gaunt, witch of a woman, with straggly red hair and long fingernails with which she would pierce small boy's arms when she had cause to admonish them for misbehaviour. I lasted one term and my school report at the time refers to my constant fighting with other boys. The option of returning to a day school in Whitley Bay seemed not to be a practical one, owing to my mother's declining health. Unbeknown to me she was suffering from breast cancer as well as severe depression and she had little time left. Thus it was to Mowden Hall near Stocksfield in Northumberland that I went next as a rather unruly raw recruit.

Mowden Hall is a fine Georgian mansion built by the Joicey family who had made their fortune out of private ownership of local coalmines. The house had been purchased by Frank Marchbank and his wife a few years earlier and converted into a preparatory school for the young sons of socially ambitious families throughout the county. Most of them were destined for the major public schools of the day, such as Uppingham, Oundle and Sedbergh. Durham seemed to be in the second league.

It was a fairly tough regime and I found it difficult to settle down, especially as my mother died in the middle of the first term. Discipline was strict and beatings by the headmaster were part of the curriculum if work fell below standard. Such treatment is unheard of today, but then it certainly focused the mind.

On the positive side the school stood in the middle of an impressive estate with woods, a lake and parklands over which boys were free to explore. There was much emphasis on sport: cricket and athletics in the summer and rugby in the winter. Although small for my age I was determined and fortunately seemed to have some innate ability, eventually playing in the school first teams. Academically I was at best 'middle of the road' and inclined to try and get away with as little as possible, hence my visits to the headmaster's study were fairly frequent.

Overall I look back at my time at Mowden with some satisfaction. I developed a sense of independence born of the need to survive. I had to learn to live with people and understand the dynamics of interpersonal skills. I enjoyed all forms of competitive sport and experienced the thrills of winning as well as the disappointment of losing. Some who know me well might also say that this must have been a time of emotional turbulence and in later life the consequences occasionally manifest themselves in a tendency to 'bottle up' personal feelings.

Recollections

I re-visited Holmslyn in 2011, nearly sixty-seven years after we moved in. I had previously written to the 'occupant' explaining I had moved into the house with my parents before the end of the war and I wondered if I could pop in and have a look to help jog my memory. In fact, Beryl Millar and her late husband had bought the house from my father when he sold it in the late 1950s. She had lived there for over fifty years. Apart from very different décor, the house and garden had hardly changed. The school next door had been replaced by flats but the open fields beyond the back of the garden had become a children's playground, so quiet residential seclusion remained ever distant. The bells to summon the servants and the master display board in the kitchen were still there.

This visit did spark a few memories of the kind which may amuse my grandchildren. A keen gardener, my father had planted a number of fruit trees in the garden. Although apples and plums were in abundance, it was some time before a favourite pear tree bore any fruit. Eventually the tree produced one splendid pear and although I was forbidden to touch it let alone pick it, a large bite mysteriously appeared on the fruit still attached to the tree.

Another oft repeated family story concerned an unfortunate white mouse. I had procured, unknown to anyone else in the family, a couple of white mice from the local pet shop. I planned to build a cage and encourage them to reproduce in vast numbers so that I could sell the offspring on to various friends. During my negotiations with the pet shop owner one of his charges disappeared. The next morning we were off for a day's shooting, when suddenly my father slapped on the brakes, his pipe fell from his mouth scattering ashes all over his trousers and he yelled, 'Bloody hell, we've got mice in the car'. Apparently the unfortunate rodent had somehow escaped from the raincoat I had worn the previous day (which now lay on the back seat of the car) having disappeared into the coat lining when it lay on the counter of the aforementioned pet shop. Amused my father was not.

A room in the house halfway up the stairs served as my playroom at Holmslyn for many years. The window overlooked a flat roof about a foot below the sill and onto which I often climbed in search of adventure. For some unfathomable reason I decided to light a small fire on this roof one day to burn some old rubbish. It wasn't a big deal and everything was undetected and cleared away. What I had overlooked, however, was that lead tends to melt when it gets hot. The leaking roof soon revealed the cause and the unmistakeable evidence of pyrotechnic activity. Another bad day. The lead patch on the flat roof was still clearly visible when I looked out of this particular window over sixty years later.

Following my mother's death, my father saw an urgent need to find someone to fill her role in the home. Various attempts included my maternal grandmother and an aunt, both of whom soon realised that such a responsibility was not for them. Eventually through a friend my father learned of a lady near Thirsk in Yorkshire who might be interested in the job of housekeeper. Lillian Peacock was about forty-five and had been brought up on a farm. She had been running a small poultry business with her sister and wanted to try something else. A visit was arranged and my father and I duly visited Lillian, who at the time was staying with her brother and sister. In later years she recalled that her mind was made up when, at first sight, I so obviously fell in love with the cat. Within a couple of months Lillian had moved to live with us at Whitley Bay. She was perfect in every way. Having lost her fiancée in the First World War, she had much love to give and I quickly became her adopted nine-year son.

CHAPTER TWO

The Fifties

POST-WAR BRITAIN is often referred to as the 'years of austerity'. Despite the introduction of the Welfare State and the nationalisation of many of the remaining major industries, recovery from years of war was painfully slow. Jobs were scarce, food and petrol were still rationed and many would have preferred to stay in the army with their mates rather than have to find gainful employment. Only one in seven families had a car, and smog was still the curse of big city dwelling. Wars also continued to feature with British troops in Korea and then Egypt as Nasser nationalised the Suez Canal. Life, however, slowly began to get better as the country began to rebuild and invest in the longer-term future. By 1957 Harold Macmillan even had the temerity to suggest that most people 'have never had it so good'. Little of this had much effect on me; I was immersed in the narrow and protected life of the educational system.

Durham School

Durham, although it might then have been classified as a minor public school for boys, has an ancient history and a proud tradition. Originally situated in buildings on Palace Green directly within the shadow of Durham's Norman cathedral, the school moved to a new site on the other side of the river in the mid-nineteenth century. The control of the school, however, remained with the Dean and Chapter and although there was an appointed Board of Governors, the influence of the Church in the running and development of the school has been a major and sometimes limiting factor over the past 150 years.

One feature was the appointment of a long series of enthusiastic headmasters with their varying attitudes towards the role that the Church should play within an educational establishment. Some were reformers, others traditionalists, but all were determined to see that the school community conformed to a Christian ethic. The headmaster in my day was Canon H.K. Luce, a tall rather thin man (his nickname was 'stick') with a pronounced hunch of the shoulders. His voice was a deep, somewhat doleful drone, which was mimicked unmercifully by the boys. He was not a sports enthusiast, although he reluctantly stalked the touchline at school matches in the winter and sat, or rather sprawled, in a deckchair on the cricket boundary during the summer. He taught Divinity to the senior boys, a subject for which I had little interest. A friend of mine (later to become a clergyman), however, drafted out my essays, which resulted in the remarkable grade of A+ in the term report – so much so that my father even wondered for a brief moment whether this subject was to be my calling.

Religious worship included a daily service in the school chapel and evensong in the Cathedral every first Sunday of the month. One of my most evocative memories of those days was standing in the gathering gloom of a winter afternoon listening to the sounds of the choir echoing around the vast arches of the Cathedral.

The opening lines of Sir Walter Scott's famous poem provides a vivid picture of this impressive building:

> Grey towers of Durham
> Yet well I love thy mixed and massive piles
> Half church of God half castle 'gainst the Scot . . .

Whilst I have not retained much religious fervour, I have a great affection for choral music.

Durham's pupils in the 1950s came from the post-war, burgeoning middle classes intent on their sons gaining academic success and sporting opportunity. The curriculum was traditional with science or classics as an option at Advanced Level of School Certificate. Latin was compulsory as an entry qualification to the Oxbridge Colleges in those days – a subject with which I had great difficulty. Whilst a handful went on to Oxford or Cambridge, the majority of those staying on for the sixth form progressed to the more prosaic red-brick universities such as Bristol, Reading or Nottingham. Also a substantial number went to Kings College, Newcastle, part of the University of Durham at that time.

I went to Durham School in September 1950, shortly after my thirteenth birthday and on the basis of my Common Entrance examination results I was placed in the Lower Remove. This was the lowest form in the school, from where the only direction was up. My first term's report was, however, dire. It was going to be a long hard struggle and my father had grave doubts as to whether I would be able to cope with the demands of the classroom.

There were three boarding houses in the school, each with about eighty boys between the ages of thirteen and eighteen. There were also about a dozen dayboys who, although attached to a house, were enviously observed returning home each evening. My house was Poole, named after a celebrated former master. The regime was fairly rigorous by today's standards. The daily routine began with a wake-up bell at 7.00 a.m. followed by breakfast in the house and Chapel at 9.00. This imposing building overlooked the school at the top of a hill surmounted by steps, commemorating the lives of former pupils who fell in the Great War. Being late for the roll-call occasioned an hour's detention. There also existed a challenge for the ambitious Pooleite athlete. This entailed sprinting from the house at the first chime from the Cathedral, careering down to school across a main road, up the steps and reporting for roll-call at the top of Chapel Hill by the ninth stroke of the hour. Not many made it. I never tried.

Lessons took most of the morning with games immediately after lunch in the winter followed by more lessons before supper. In the summer with extended daylight the lessons came before the games. Prep each evening for a couple of hours was followed by House Prayers and then bed. There were four large dormitories in the house each containing about fifteen beds of the iron bedstead variety. There was minimal central heating, windows were wide open whatever the weather, and there was no privacy whatsoever.

My brother had left at the end of the term previous to my going to the school and as is so often the case, I not only inherited his nickname, Dex, but also experienced a variety of reactions from boys who had known him. At the senior end there remained some of his close friends, many of whom I had already met. They were anxious to show no favouritism towards me but at the same time I was aware that they kept a brotherly eye out for potential problems. At the other end of the scale there were the inevitable few who had cause to make my life uncomfortable, perhaps as a result of their being on the receiving end of my brother's displeasure. Such is the pecking order in boarding school institutions.

Derek had been a competent rugby player, known for his fearless tackling rather than any great ball skill in open play. More importantly he had joined the Boat Club in his first summer term as he found the alternative of cricket beyond his powers of concentration. Small for his age, he began as a cox and soon found himself steering the first crew in their pursuit of glory on the River Wear. He eventually began to row himself and finished up Captain of Boats with all the schoolboy glory that exalted position carried. Thus, at least on the sporting field, there developed an expectation from those around me that I should follow in his footsteps.

Durham had a proud sporting tradition. Rugby was the winter game played by everyone whether they liked it or not. When it became obvious that a boy was a hopeless case he was allowed to choose an alternative and could go for a run – either Farewell or Houghall – about three or four miles. Fives was also an option but everyone had to do something unless they had a note from Matron that they were ill. Having had the benefit of playing rugby at prep school this was an opportunity where I was keen to shine and soon got into the Little Clubs side, again the bottom rung on a very long ladder. A number of boys had never played the game before and on one occasion one such caught the ball on his goal line and proceeded to swerve and sidestep the opposition before arriving at the opposing goalposts where he promptly threw the ball over the crossbar!

In the summer term boys could choose between cricket or rowing. I rather fancied myself as a cricketer, having been in the Mowden First XI and reasonably capable with the bat. I therefore elected for cricket, somewhat to the surprise of some of my brother's contemporaries in the senior school. As he had been an oarsman of some repute, his younger brother was expected to follow in his footsteps. My success on the cricket field, however, was short lived and I duly converted to the water in my second year.

At the age of sixteen I gained sufficient 'O' Levels to warrant my elevation to the sixth form and chose to take science – biology and chemistry – at 'A' Level. I hadn't the slightest idea what I wanted to do other than go on to university so that I might continue to row or play rugby. Life in the senior part of the school was infinitely more enjoyable than in the early years. I coped tolerably well with the work and succeeded in getting into the First XV for three consecutive years. It happened to be an excellent final year, in which the side were unbeaten. Our fly half was Mike Weston, a gifted sportsman who subsequently played over thirty times for England. In the summer I became a dedicated oarsman, progressing from junior to senior status. In the summer of 1954, the first crew won all the senior events in the north-east, a feat never achieved before in the school's history. I was in the second crew that year and we also recorded some resounding victories. The following year I emulated my brother's appointment as Captain of School Boats.

I have many vivid memories of those first couple of years at Durham. The time was filled completely either in the classroom, on the playing field or on the river. In addition there were lots of other activities, clubs and societies to soak up the spare time. The archaic tradition of fagging for senior boys was in operation as was beating by the School Monitors for the inevitable misdemeanours. It was a fairly tough regime and there were casualties. Survival is perhaps an over emotive term, but you had to get on with it. There was little sympathy or understanding for the faint-hearted.

Endless debates continue on the merits of a boarding school education. The obvious advantages ought to include a higher level of academic teaching with smaller classes and more individual attention. Teaching standards at Durham were as variable as they are in most institutions and whether I would have achieved the necessary exam results to go to university had I been taught in the state sector is anyone's guess. What is certain, however, is that the opportunity to become involved in competitive sport was infinitely greater and that alone has been a marked influence in my life. This is not simply from the aspect of striving for success on the field or on the river, but from the experience of teamwork, which most sports require of their participants. For me this has led directly to friendships which have lasted a lifetime. Boarding schools are very different nowadays. Most are co-educational, much friendlier establishments without the unhealthy emphasis on senior boy autocracy. But there remains for the child the challenge of living away from home and learning the lesson of having to get on with people within a community. On balance and always provided that the parents can afford the fees without sacrificing family lifestyles too much, then I think I would still favour such an education. For most of their secondary education we chose this option for our own children, but much had to do with the alternatives (or lack of them) available in our locality at the time.

A Fresher at Kings College

Looking back it is interesting to reflect on the reasons that I chose to follow my brother to Kings College and the Department of Agriculture. They certainly were not based on a deep ambition to get into practical farming, even if there had been sufficient money to so embark. Probably the main driver was the desire to avoid National Service, which was still a compulsory obligation for every male over the age of eighteen in the UK in 1956. This was a two-year stint in Her Majesty's Armed Forces and unless you were successful in an application for deferment due to further education, in you went. My brother, having completed his degree, was by then in the Royal Artillery stationed in Germany, with the unlikely rank of Lance Bombardier. His reflections of the life and the square-bashing held little appeal for me. But first I had to be accepted as a student.

Apart from the 'A' Levels, entry into the Faculty of Agriculture required at least a year's practical farm experience. In fact a high proportion of the students were either farmers' sons or they had already been in the army as well as completing a year on a farm. My experience had been confined to various school holiday jobs on a mixed arable and stock farm near Earsdon about a five-mile cycle ride from home in Whitley Bay. My father knew someone who had a contact within the National Coal Board which owned all the surrounding land. This unlikely arrangement was due to the widespread coal mining in the area and the resulting subsidence. Owning the land was cheaper for the Coal Board than paying out compensation to the original farmers.

One might have hoped that an organisation such as this would have employed modern technology in their farming enterprise, but the main source of power was horses not tractors. Most of the staff were ex-miners and the work was straightforward labouring with virtually little opportunity to learn anything. Such exciting activities as mucking out calf boxes, endless days singling turnips, building haystacks and stooking thistle infested - oats, hardly contributed to the total sum of my meagre knowledge. Nevertheless it was my first job and the feeling of hard-earned cash at the end of the week was rewarding.

Another fortnight's holiday job was down near Thirsk in Yorkshire. This was Lillian's brother's farm, again a stock and arable enterprise. Brother George was a dour man, seeing me as a source of cheap labour rather than an enthusiastic student eager to learn how he managed to make money out of his business. My stay was marked by an unfortunate incident. George asked me to remove and burn the loose straw covering his store of potatoes in one corner of a field some distance from the farm. This was prior to riddling and bagging the crop for sale. It seemed to me much easier to light the straw in situ rather than remove it as there seemed to be a protective layer of soil underneath. What I hadn't realised was that there was more straw under the soil protecting the crop from frost. Naturally enough the flames engulfed the second layer of straw and at the same time roasted the outside layer of potatoes. George was not a happy bunny. But he was an enterprising old devil and managed to claim full insurance as well as sell the aforesaid 'roasties' to a neighbour for stock feed. I didn't bother to ask him for a reference.

I attended an interview at Kings College in my final term at school and I remember being ushered into the presence of the Dean, Professor Mac Cooper. I knew I was on thin ice if it came to an inquisition of my scant knowledge of practical agriculture. I also knew that if I failed to get in I would be consigned to spend two years in the army and then to have to complete a full year on a farm. Indeed had this been the outcome it is unlikely that I would have returned to agriculture, perhaps seeking a slot in the family business. My interview, however, turned surprisingly quickly to matters of sport. It appeared that the good Professor was a keen rugby man and was interested to hear that I had played for the Northumberland schoolboy team. The rest of the time was taken up with a discussion on the merits of rowing as opposed to rugby and before I knew it the interview was over. Little did either of us know at the time the portent of this first meeting.

I was delighted to receive the news that I had been accepted and that I should commence at Kings College as an undergraduate in October 1956.

Kings College, formerly Armstrong College, named after the famous late nineteenth-century industrialist, Sir W.G. Armstrong, was situated in the centre of the city of Newcastle. The Department of Agriculture had been founded in 1891 and had built up a formidable reputation in early research programmes focusing on means of increasing productivity through improved breeding of livestock and new techniques in grassland management. Progress since the war had, however, slowed to a point of stagnation and Cooper had been recruited in 1954 to galvanise the Faculty. Although at the time only a distant figure to us lowly undergraduates, it was clear that this controversial New Zealander was in the process of sweeping away the old traditional practices as well as many of the staff who had been coasting along for years and were well past their sell-by date. Mac Cooper was to become one of the leading figures in British agriculture over the next twenty years.

There were about thirty of us in the year's intake, many of whom were three or four years older than me, having either completed National Service or been working on farms before coming up to university. At barely nineteen and woefully ignorant of agricultural terminology, I attempted to keep a fairly low profile amongst my fellow contemporaries. In fact our first year was virtually a repeat of science 'A' Levels and we joined with other faculties such as medics and pure scientists for lectures and laboratory work. Only a handful of lectures were within the Agricultural Department, although each Wednesday we would have a field day visiting farms in surrounding Northumberland.

As my father and Lillian were still living in Whitley Bay, I was able to live at home and travel in on the train to college each morning. This arrangement had many obvious advantages such as creature comforts and no cost. Neither did it adversely affect my involvement with student activities and sport as my father had purchased for me a car – an untold luxury for a student in those days. When Harold, my father's younger brother, joined the family business after his war service in the RAF, they decided to diversify into the motor car trade. By the early 1950s they owned two garages in the vicinity of Whitley Bay (Wilsons in Monkseaton and The Coast Road Motor Company in North Shields). Harold ran them both successfully and one obvious bonus was the availability of cars for all the members of the family at cost price. Mine was a Morris Minor Tourer with the never-to-be-forgotten registration number of HTY 674. Furthermore it was brand new – not one of those old bangers which would never start. There were, however, some strict rules in its use. It had to be kept in the company garage about a mile walk from the house. I was to use the train to college unless needing the car to go to sporting venues and although there were no breathalyzers in those days, my father was keen to impress upon me that there should be no drinking and driving. I regret to say that there were subsequently numerous occasions when I found myself behind the wheel when I clearly should not have been.

Most of my contemporaries were in digs or in one of the halls of residence. The former option was either a house or a flat in one of the less desirable areas of the city. I envied their independence, but not their living conditions, often in damp, cold rooms with the barest of furnishings. Many of my friends who have survived from those days look back with gratitude to the invitations, which they received when they were impecunious students, to spend Sundays at our home in Whitley Bay. My father had always been an enthusiastic supporter of my sporting activities, either rugby or rowing, initially at school and now at college. He thus began to meet a lot of my friends and thoroughly enjoyed being with them. They always received a warm welcome as well as outstanding food produced in vast quantities by Lillian.

On a number of occasions we played golf with my father. He and Harold were well known in the town and were keen members of the Rotary Club and for a while the Masons (however, for some reason my father fell out with this secret society in later years and resigned) as well as stalwarts of Whitley Bay Golf Club. We students were 'hacker golfers' and it was with some risk that my father signed us in on a Sunday morning, hoping no doubt that we would not cause too many raised eyebrows on the course. One day I informed him that one of my friends, Ralph Dawson, was to join us for a game. Ralph had grown up in Kenya and had arrived at Kings to study civil engineering. In a moment of weakness he had admitted to playing in the Kenya Boys Open Championship. By the time the story had been embellished somewhat by his friends, Ralph was reputed to have actually won the event. Father, hearing of this reputation, summoned his friends to the first tee to observe the hapless Ralph drive forth. Three air shots later and a swing which rather gave the game away, the disillusioned crowd dispersed quickly back to the bar to discuss the merits of golf in the colonies. Ralph never lived it down and to this day is reminded of this coveted award.

The general agricultural degree at Newcastle was supposed to have a broad base and a practical emphasis. Accordingly it attracted a high proportion of farmers' sons who intended to return home after graduation following the three-year course. As mentioned, the first year was heavily oriented towards pure science to ensure that everyone knew the basics in zoology, botany and chemistry prior to opening up the subjects of animal and crop production in the subsequent years. Some struggled in that first year, especially a mature student (in his late twenties) called Jimmy Egan. Jim had been in the regular army serving in Malaya and coming from a harsh mining background in Durham had only a meagre knowledge of the sciences. How he managed to gain entry was always something of a mystery to his contemporaries, although the Prof was known to make exceptions for individuals who showed flair and personality. Jim certainly had these suits in spades. In the first year practical chemistry exam there was a question which required the technique of a titration to be performed. In this case the two liquids were potassium permanganate and arsenic oxide. The idea was to use what was called a pipette to suck up a measured volume of one of these liquids into a conical flask and put the other into a vertical tube (burette) up which ran a graduated scale. At the bottom of the burette there was a tap by which the volume of liquid dispensed into the conical flask could be accurately measured. At the point of colour change in the conical flask the scale reading could be used to work out the answer to the question posed on the exam paper. In this particular case, it became clear that no one in the lab. knew whether to put the colourless arsenic oxide into the flask or the purple permanganate into the burette. Never slow in decision-making, Jim proceeded to advertise his intention by filling the tube with purple, thus indicating the next job was to suck up the arsenic oxide. Before long most of the students were following his example and the lady in charge had to stop everyone in their tracks before someone poisoned themselves.

Jim was a leading light in many ways and although he failed his first year and had to repeat it, he eventually completed his degree and emigrated to Canada where he had a successful career within agricultural land development and reform.

The second year student

The second year at college involved a more relevant curriculum to an agricultural course and in consequence proved much more interesting. The average day would involve perhaps three or four lectures in the morning with laboratory work in the afternoon. Field days were a welcome distraction with visits to the Faculty's research farms at Cockle Park or Nafferton and also to local commercial farms. There were also some surprising projects, for example, collecting and pressing wildflowers, grasses and weeds. Four of us would set forth to a chosen site (well away from public gaze) and gather all manner of vegetation, placing it in a large cardboard box, which we then transported to a nearby pub. After a few beers we would then proceed to attempt to identify the contents of the box into the various species and distribute samples between us. The end result had to be submitted for approval to the Head of the Agricultural Botany Department, a large kindly woman, by the name of Dolly Clark. Fortunately she could see that the dedication and skills required to present an album of carefully pressed specimens was somewhat beyond the capability of some of her less gifted students.

The general standard of lecturing was low. There were, of course, some notable exceptions, not least the Dean himself, Prof Cooper. He was a tall rather gangling individual with a craggy face and somewhat unruly hair. With a pronounced New Zealand accent and a reputation which went before him, most of us felt compelled to turn up to his lectures. Unlike other of his colleagues, he used no notes at all and tended to sit on the table at the front of the lecture theatre, swinging his legs and talking to the audience in a matter of fact sort of way. He was essentially an animal production man and a great advocate of using grassland more efficiently as a means to improving economic performance. Indeed by the time he came to Newcastle he was already a well-known name in agricultural circles. From a modest farming background in the North Island of New Zealand, he had come to Britain as a Rhodes Scholar at Oxford. Although not distinguishing himself academically, he excelled at rugby football and played in three successive 'Varsity matches as well as being selected for Scotland, the country of his grandparent's birth. He married a girl from Oxford and after a spell back in New Zealand and war service in Italy, he returned to the UK in 1947 as Professor of Agriculture at Wye College in Kent. This was post-war Britain and food rationing. Mac Cooper recognised that farming practice was in dire straits and he lost no time in informing both Government and farmers that in his view, 'British Agriculture was running at half cock'. He welcomed the controversy that followed this criticism and he took every opportunity to challenge the traditional attitudes which prevailed throughout the 1950s. So it was perhaps unsurprising that we band of undisciplined undergraduates sat politely listening to the great man extolling the virtues of Jersey cows, Italian ryegrass and Wild White clover.

At the other end of the scale we had totally incompetent lecturers. One such was a small, bespectacled man by the name of Jack Blake. His subject was Agricultural Economics, at best, a dry mixture of theory and mathematics. In Blake's hands it was a morass of indigestible rhetoric, which he read verbatim from a set of notes in a dull monotone throughout the period. The man had, amongst other disabilities, a hearing impediment. At the start of the lecture he was apt to switch off his hearing aid in case anyone was unwise enough to ask a question. On one famous occasion we decided to tape-record part of the lecture. On playing it back all that could be heard was a series of loud conversations like those taking place in an audience before the Chairman has called the meeting to order, interspersed with loud bangs as people burst balloons. If one listened carefully, one could just distinguish Blake's voice droning on about the Law of Diminishing Marginal Returns.

This subject is the only one where, in the terminal examination, I actually changed the answers I had written to the questions. It seemed on re-reading; that the answer I had written to question two was more convincing than the one I had written to question one. In fact one of our numbers, Maurice Donnelly, actually scored no marks at all. Blake took great delight in telling him in front of all his colleagues that he couldn't even give him a mark for spelling his name right because he (Blake) didn't know whether there were two n's or two l's in Donnelly!

Although fifty years on, this all sounds like horseplay, but the importance of passing the annual examinations determined whether you continued with the course. Some were casualties and either had to re-sit, repeat the whole year or simply leave and find a job. This meant, as the dreaded time approached, feverish activity in the library and the notes of the swots who had attended all the lectures were in great demand. Fortunately most of us made the grade, albeit with little to spare, and the annual results heralded the aptly named Joy Week, where towards the end of June we celebrated as if there was to be no tomorrow. We played golf, we swam in the sea, we drank copious quantities of beer (at what in today's currency amounted to 5p per pint) and we talked late into the night about the future. These were formative years where lifelong friendships were formed and the sun seemed to shine every day.

The three-month summer vacations meant, for most of us, finding a job, either on a farm or building site. With the asset of a car, I was in the enviable position to organise a holiday, usually in September, for four of us to go off and see some of the world. The first of these trips in the summer of 1957 was to Austria via the cross-Channel ferry to France and down through Germany. In fact we managed to visit Venice as well for a couple of days and at the end of the four-week trip, the International Exhibition in Belgium which featured the famous Atomium. These were low cost adventures (£25 each would see us through the entire trip) with all our belongings crammed into the boot of HTY, together with a tent tied to the boot lid. I was the only one insured to drive and we proceeded at a modest 50 m.p.h. for over 2,500 miles. It was a marvellous experience and one I repeated in the next two years with different compatriots to include Spain and Italy.

Rowing

My memory of those first two years as a student revolves around the sporting and social life of the college. Although I attended lectures I found the work uninspiring as I had already covered most of it at school. Probably the most difficult decision I had to face was whether to play rugby in that first winter or join the rowing club. It simply wasn't possible to do both as the rowers were fully involved in the winter as well as the summer. In the event I chose to row, mostly because I already knew many of those in the Boat Club having competed against them the previous summer. Kings College Boat Club was at Newburn on the River Tyne, about four miles up river from the centre of Newcastle. I was familiar with the set-up but that had been during the summer. Winter rowing was altogether a less enjoyable experience. Firstly the river was tidal, so that the landing stage had a series of steep steps and three separate platforms which were used to launch boats. As the tide went out it left a thick layer of filthy mud covering the whole area which made it both lethal, slippery and disgustingly smelly. The Tyne in those days was little better than an open sewer with unspeakable objects floating along on the surface, some of which landed in the boat or splashed on to your back if the wind was blowing against the tide. To add to the pleasure the weather was usually foul and our outings were on a Sunday morning following a heavy night in the bar or on the town. Dedication was the watchword, but youth knows little better.

Although we used the Kings College Boat Club, the crew I was in included students from the Medical Faculty and some from the colleges in Durham. In fact, it became the Durham University First VIII and as such we were set on a course of serious coaching and preparation for the premier winter events such as the North of England Head of the River in Chester and the Thames Head of the River in London. The latter was also the University Athletic Union Championship for all the English universities with the exception of Oxford and Cambridge.

Head of the River races follow a set pattern. Each crew is allotted a starting number based on their club's finishing position the previous year. Rather than race abreast, the crews start at fifteen-second intervals and follow each other down the course with the objective of recording the fastest time to the finish. The better crews tend to start first so at the front of the race there isn't usually much congestion. As more crews commence, however, and greater variation in ability becomes evident, the faster boats start to overtake the slower ones. As a rower you are of course facing backwards and only aware of the opposing crews to your stern, either as they recede or as they begin to reduce the distance of water between you. The cox on the other hand can only really see the activity in front and has the unenviable task of steering the boat to take best advantage of the tide or current as well as the bends in the river. By the middle of the race there might be two or three eights racing down the middle of the river, all with their cox shouting obscenities at each other to give way. Meanwhile the rowers strive to keep in time and find the reserves of energy to propel the boat full tilt towards the finishing line.

Undoubtedly the short straw is held by the cox who transpires to be either hero or villain depending on whether he has steered successfully through the chaos or clashed oars with another crew and managed to get the crew disqualified. Either way he is likely to end up in the river as tradition has it that such a fate is part of rowing folklore.

The North of England Head and the Thames Head were scheduled for successive weekends in March every year. The Durham University crew travelled to Chester in whatever transport was available. My car was by far the most reliable and our stroke, Robert Jackson, a dental student, also had a fairly robust little car. The other vehicle was an old van belonging to our number two, Jimmy Nicholson. Jim was a mature student studying electrical engineering at Kings. He was in his mid-thirties and had a small business in Bedlington not far from Newcastle. He was a broad Geordie and virtually impossible to understand if you happened to be from another part of the country. His van was in a terminal state of disrepair and he carried with him an assortment of tools to deal with problems as they arose. The back of the van was also permanently full of decaying TV sets and parts of washing machines belonging to his long-suffering customers, so that besides himself he could only transport one other member of the crew. The final part of the procession comprised a lorry fitted with a wooden superstructure designed to carry our boat and the oars. Needless to say this caravanserai took its time crossing the Pennines towards the ancient city of Chester. No motorways in those days.

We were billeted in The Queens Head in the centre of town and within easy reach of the River Dee. Little did I imagine that within a few years I would be moving to this part of the world on a permanent basis. The race had attracted about thirty crews and as the University had done well in previous years we started fairly near the front. It all went according to plan and we finished first thus qualifying for a well-earned party in the evening at the Royal Chester Clubhouse. With thick heads we then set off on the second leg of the journey to London and the River Thames.

This was to be an altogether more impressive experience. Along the towpath up-river from Putney Bridge stood numerous palatial boathouses belonging to, among others, Thames and London Rowing Clubs. This was the Mecca of British rowing and we were to have a week before the event training amongst the elite. The four-mile course was identical to that rowed by the Oxford and Cambridge crews (this race was to be held the following week) except in reverse. They start at Putney and row upstream to Mortlake, whereas the Head of the River Race started at Mortlake and finished at Putney. As the river is tidal at this point, so long as both races were timed to coincide with a following tide, it mattered not in which direction you happened to be going.

On the day of the race the whole river was seething with rowing eights. The event attracted something over 300 crews from all over the UK and some from abroad. All had a specific starting time and the sight of all these boats wending their way up-river to begin the race was an impressive spectacle. We duly arrived on time and started at number 37, the position at which Durham University finished in the previous year. I have a record in the rowing diary I kept in those years, which states:

> We got a poor start and did not settle down until after the first mile. We then began to overhaul Walton 1st VIII who were in front of us. After two miles we had passed them and were in full cry after RAF Cardington 2nd Crew who were in turn on the tail of University College Hospital 1st's. After Hammersmith Bridge it rather died for half a mile and we lost half a length of what we had gained. At Craven Steps, (no relation) however, we managed to get it up (the stroke rate that is) and at the finishing post were only half a length down on number 32, although 33 had gone past. We were all in!

> Results: 1. Isis (Oxford University 2nd crew) in 19.18 mins. 2. Goldie (Cambridge University 2nd crew) in 19.37min. 21st equal. Durham University in 20.18 mins.

The following year we repeated our success at Chester and improved our position to sixteenth on the Tideway (the rowers' terminology for the Thames) at the same time winning the University Athletic Union award.

Rowing occupied most of the summers of 1957, 58 and 59. Regattas were held on the principal rivers of the north-east, including the Tees and the Tweed as well as the Tyne and Wear. Durham was the major event spanning two days and for the senior fours included the Wharton Challenge Cup on the first day (restricted to the Durham Clubs which included the School, University and the City) and the Grand Challenge Cup on the final day (open to all-comers). Over the years either at school or at university I managed to be in a winning crew for most of the senior events at some time or another with the single exception of the Grand. In my final year we beat Durham City on the first day by a couple of lengths but on the second day we lost by 3 ft. in the final of the Grand. Many of our competitors were good friends against whom we rowed for years in succession, one of whom, Ian Shepherd of Durham City, continues to remind me of this omission in my rowing career whenever the subject is raised – usually by him. He tells me that he won it eight times.

For the most part we took the sport very seriously, training hard under the close scrutiny of a coach riding along the towpath on a bicycle. The winners usually took home a handsome pewter tankard or silver-plated cup as a prize and some of the more successful participants ended up with a cabinet full of trophies. At the end of the summer season most of the crews who had been competing against each other turned out for the regatta at Talkin Tarn, a half-mile long lake near Brampton in Cumbria. Although there was the full range of senior and junior events, the fact that for the longer course the crew had to turn round a buoy at the far end of the Tarn and row back to the start, rather turned the event into an end-of-term jamboree. More importantly, as soon as your crew was knocked out you made your way to the small pub in the village where by about 8.0 p.m. you could barely squeeze in. My brother Derek, still rowing at this time for Tyne Rowing Club was ensconced on the piano in one corner hammering out such well-known songs as 'The Wild West Show' and 'The Bastard King of England'. To get another pint it was easier to climb out of the window and get it from the pub next door. At closing time the pub emptied and by one means or another its occupants transferred to the Regatta Dance in Brampton, where on entry you were rubber stamped rather than issued with a ticket. Rugby rather than dancing seemed to be on many minds. The final chapter in this unseemly sequence involved a journey back to the Tarn, where most of the rowers had pitched camp for the night. Before turning in however, some of the braver hearts would venture out in the various pleasure boats moored at the side. Obvious targets for pitched battles and sinkings, the rivalry continued until dawn.

Two memorable tales of Talkin Tarn Regatta keep being re-told over the years by those of us who still meet and were lucky enough to be part of this scene in the late 1950s. One concerned Jimmy James, a contemporary at Durham School. He had been born in India and was due, as was his custom, to return there for his summer holidays. Jim was a fine sportsman with rather less enthusiasm for the academic side of the school curriculum than his parents had hoped for. On this occasion he had to catch the late-night train from Carlisle to London and thence the plane the next morning for India. In order not to miss the fun in the pub and the dance he decided to take the last bus from Brampton to Carlisle at 11.0 p.m. This would give him twenty minutes to catch the London train. As might be expected the evening gathered pace and it became necessary to support Jim from the dance to the bus stop. It was a matter of about 200 yards and in the days when buses had conductors as well as drivers. The sight of this little trio weaving its way up the street did little to impress the bus officials that their final passenger would be able to contain himself over the nine-mile trip to Carlisle. Arriving at the door, the conductor leaned down the steps and inquired, we thought rather aggressively, 'And where do you think you're going?' The reply was short and to the point, 'Bombay'. Amazingly he got there.

The other apocryphal tale concerns the legendary Kenya Boys Open Golf Champion, Ralph Dawson. Ralph was no rower but he and numerous other rugby types used to come up to the Regatta to join in the festivities. All went well until late in the evening he found himself in one of the pleasure boats on the Tarn and in severe danger of being boarded by friends and foes alike. With great presence of mind Ralph removed all his clothes in anticipation of ending up in the water. Sadly the boat turned over in the melee and down to the bottom went every stitch he had. What made matters worse he had to be back in Newcastle early in the morning and it was sixty miles away. It transpired that one of our more sober and respected colleagues of the time was a solicitor who also had to return to Newcastle in his rather snazzy little sports car. A polite request to give Ralph a lift was affirmed and Ralph gratefully, if self-consciously climbed into the passenger seat. Derek Baty, the driver, who had not had the pleasure of Ralph's acquaintance recalled that little conversation took place between them and he remembers bidding Ralph farewell as he walked off down Jesmond Road clad in a blanket at four in the morning.

Newts

Whatever the collective noun for a gathering of newts might be, they are now apparently, a fiercely protected species. Builders and developers must spend thousands of pounds ensuring these small, slippery, lizard-like creatures are able to maintain their habitats. Some species, however, are not so protected and have to take their chance in life's lottery. In the context of this memoir 'The Newts' are not small or slippery (at least not all of them are), rather they are the generic name for a group of people of similar age and inclination who were around in the late fifties and early sixties. One might even describe them as a dining club, but somewhat down market from the Bullingdon variety. Some have been mentioned by name already, usually because of their personal involvement in an anecdote. But the most remarkable fact is that more than fifty years on, those of us who are left still meet in Northumberland each October to relive our student days and remember our lost youth. In fact we usually convene in the summer as well and include our wives who sit patiently listening to the same old stories year after year.

There is no *raison d'être* for this happy band. The common interests were sport, drinking large quantities of beer and having fun. Some were contemporaries at school, others were competitors on the river, most were undergraduate students at Newcastle. The leading light was, and indeed still is, Tommy Nicholson. Slightly older than the average, he read classics and was a star turn when our debates focussed on the ancient capitals of Greece and Rome. Tom was also a local lad, born and bred within the smell of the Tyne. He was not a sportsman, but qualified with honours at the bar. To those who didn't know him, he could appear abrasive, even aggressive. He had an air of authority and presence. As our Honorary Secretary he organised our dinners and afterwards placated the chefs and landlords. Tom was in charge and frequently cracked the whip when wayward members let the side down. In later life he taught English at one of Gateshead's large schools but he retired early when the relaxation of disciplinary standards became too much for him to bear. The start of 2011 coincided with Tom's eightieth birthday. His health had not been good for some time and he was loath to celebrate this important anniversary. Not so his fellow Newts. We met for lunch in Durham and only a handful of members were unable to attend. He told me later that it was one of the happiest days of his life. This is not the place to describe this fine body of men and their exploits, all of whom are the wrong side of seventy. Perhaps one day I will write another book to their memory and the influence so many of them have had on my own life. A couple of years ago I suggested that perhaps we should call it a day before only a handful of us were left. Not a bit of it – last man standing takes all.

Many people, when asked what they have gained from a university education, place the acquisition of knowledge near the bottom of their list of benefits. Subjects vary of course but often a degree simply opens a door. To get through it into a career, even as a trainee, requires other assets such as personality, self-confidence, enthusiasm and judgement to name but a few. These characteristics develop as a result of being exposed to and immersed in life's experiences. As a student these should be many and varied. This is the opportunity to test ideas, challenge the status quo, justify prejudices and formulate values. Contemporaries are doing the same and because their age and experience is similar to your own, you are more likely to listen to their opinions. This is why these formative years are so well remembered and the friends you make are often retained for life.

What next?

Towards the end of my final year at Kings College in 1959, the question arose as to what to do next after this glorious three-year spell of student life. Why could it not continue forever? In fact, the issue of National Service played a major part in the decision for the second time. It was evident that conscription was about to be discontinued within the near future. By opting to undertake a two-year postgraduate study for an MSc, I could further defer my call-up and maybe avoid National Service altogether. It all rather depended on achieving a satisfactory grade in my final exams. In the event I was awarded a second class honours, although in the lower division – just sufficient to convince the authorities to agree to my continued study.

So at the age of twenty-three I was destined to become a postgraduate student for the next two years. Further than that I had no idea of what I was going to do or what sort of career I was going to pursue – something would probably turn up. Such were the irresponsibilities of youth.

CHAPTER THREE

The Early Sixties

THE SIXTIES WERE A REMARKABLE DECADE. International events such as the Cuban missile crisis, which almost plunged the world into nuclear war; the assassinations of John F. Kennedy, his brother Robert and Martin Luther King; the Profumo affair; the arrival of the Beatles; winning the football World Cup as well as a social upheaval of spectacular proportions as the contraceptive pill introduced 'free love'. Finally in 1969 man made the giant leap for mankind onto the moon.

For me, the first part of the decade was the time to make a giant leap into the unknown – a transition from student to worthwhile employee. It was also the time to accept the challenge and the exciting prospect of marriage and fatherhood. But there were still a couple of years in which to pursue my somewhat irresponsible life.

Postgraduate life

I had little idea what my extended stay at University involved, other than it would be very different from the life I had been leading as an undergraduate. My interview with the Dean revolved around the broad area of research that I would undertake. He suggested some work involving the evaluation of conserved grass silage using sheep as a means to assess the relationship between stage of grass growth, the effect on appetite and consequent live weight gain. This sounded interesting and meant that I would live and work up at Cockle Park, the University research farm, near Morpeth, about fifteen miles north of Newcastle.

Life at Cockle Park was fairly basic, but hugely enjoyable. There were about ten postgraduate students and a couple of people who had lectureships in the Department as well as being engaged in research. We all lived in the old Pele Tower, an edifice built of solid stone, within which in the fifteenth century the local inhabitants as well as their livestock took refuge when the sheep and cattle rustlers arrived. Little had been done over the years to make it into a welcoming student hostel. In the summer life was tolerable. In winter, however, the notorious north-east wind howled through the gaps between the stonework and window frames with the inevitable consequence that most of its inmates spent many a dark evening in The Oak, a pleasant little pub about a couple of miles up the A1. The landlord was an expert darts player and could be relied upon to finish a game with a double without leaving his position behind the bar counter.

I suppose I assumed that there would be some facilities with which to undertake the research work which had been agreed, but it soon became evident that I was expected, with the help of farm staff, to construct them myself. I have always enjoyed building things, maybe an inherited trait from my grandfather, so within a few months fourteen experimental silos were erected and work started to convert the inside of an old building into pens for twenty-four sheep.

Conserving grass for feed in winter is a vitally important part of a livestock farming system. The traditional method had for years been hay, but this is a notoriously risky operation as the British weather is so unpredictable. In the fifties silage-making had begun to take off and there was great interest in the various techniques and machines which were coming on to the market as farmers and researchers looked for the most efficient systems. Silage-making is essentially a fermentation process. The idea is to make a big heap of grass and to create conditions within it that encourage the right sort of bacteria to proliferate and in so doing produce lactic acid which pickles the grass and prevents its decomposition into compost. One of the big problems is wastage. Whether in a big open clamp or inside in a building, it is essential to consolidate the cut grass in order to squeeze out as much air as possible. If this is not achieved, the material overheats and nutrients are lost as heat and carbon dioxide. At the other extreme, especially in wet conditions, the clamp can become virtually waterlogged and another sort of bacteria predominates, producing butyric acid, resulting in an evil smelling mess that the animals won't eat.

My research project was geared to exploring the effect of various treatments on the losses during conservation, mainly related to the maturity of the cut grass, but also the use of chemical additives. During the winter, by feeding the experimental silage to groups of sheep, I could also measure appetite and live weight change so that the overall difference between the treatments could be evaluated. The work involved a massive collection of physical data as well as chemical analysis of grass, silage and even sheep faeces.

In the event the first two years went very well and I was invited to give a seminar to staff and researchers in the Department. The results were so encouraging that I was invited to stay on for a third year and to convert my postgraduate degree into a PhD. I jumped at the chance. The end result for those with a thirst for knowledge is contained in my PhD thesis, entitled, grandly and obscurely, 'The Value of Silage in Sheep Nutrition'. I suspect only four people have ever read it: Prof; the external examiner (an eminent professor from Edinburgh whose name I can't remember); the typist and me.

Rugby football

Virtually all my contemporaries had left at the end of 1959 after graduating, including those who were in different faculties and who had been my rowing friends. It therefore seemed opportune to try another sport, so I joined Kings College Rugby Club, anticipating that there would be little problem finding time to play the game on Saturday and Wednesday afternoons. I also re-joined Northern Rugby Club, where I had played schoolboy rugby in the holidays.

The Kings College 1st XV in those days was a formidable outfit, playing the senior clubs throughout Northumberland and Durham. Kings ran five or six sides and competition to get into the first team was intense, as this was the essential initial step towards playing for the University. As with rowing, the University side played under the aegis of Durham and was a mixture of Kings College, medics (who had their own separate club) and the individual colleges in Durham itself. The major competition was the University Athletic Union (UAU) Championship, within which Durham had an impressive record. This was organised on a regional basis with our main adversaries being Leeds, Sheffield, Hull, Manchester and Nottingham.

The three-year lay off from rugby must have done me good and within a couple of trial games I was in the Kings 1st XV and the University side. Apart from the challenge of the game, my involvement opened up a whole new range of social contacts. Some had already played for county sides (Brian Stoneman for Durham and Mike Hymas for Northumberland), also Derek Morgan, who as a dental student in the University side, was to be selected to play for England later in the year. Derek, a mobile and powerful No 8 hailed from South Wales. Having played for the Welsh Secondary Schools, he was advised to apply to the Dental School in Newcastle, as the Dean was a rugby-mad Welshman. Legend has it that he was the only student ever to be accepted into the School without first having been interviewed by the Dean. When it was realised that Derek had been born in Monmouthshire and therefore qualified to play for England, the chances of his succeeding in his dental career received a severe setback! Nevertheless, the England selectors went for

Derek and he played about eight games for England before he badly injured a knee. After his time at Newcastle he continued as an administrator in the game, principally involved with student rugby and as England Manager on a number of overseas tours. In 2002/03 he was elected President of the Rugby Football Union.

We played most of the senior clubs in Northumberland and Durham on Saturdays during term time and University games on Wednesdays. Training sessions were two evenings a week and sometimes on a Sunday morning. So it was serious rugby coupled with heavy nights in the Bun Room – the College Union main bar. In my second season I was elected Captain of both the Kings College Club and the University sides. In the vacations I played for Northern, in those days with Gosforth – their neighbours on opposite sides of the A1 – the two leading clubs in the north-east.

In terms of results, I can remember nothing unduly noteworthy. No County Cup wins and no real progress beyond the regional group in the UAU Championship, despite having a side which boasted about eight or nine county players. More memorable were a number of tours, one with the College side to the Manchester area and the other with the University to the far south-west. Nowadays when school sides go as far afield as Australia and New Zealand on tour, such trips appear tame indeed. But for many of us a tour (especially as the University paid for transport and accommodation) to Cornwall was equivalent to a foreign country. Rugby in that part of the world was a serious business. We were advertised as a 'star-studded' side, probably due to Derek Morgan's reputation, although he wasn't on the tour. The consequence was a turnout of about 5,000 spectators at each of our matches against Penzance and Redruth, all baying for blood. We might even have won the games, I can't remember, but with the exception of our hotel manager, they seemed pleased to see us and gave us a wonderful time.

One other highlight, which came my way as a Durham University player, was selection in the British University's XV to play the Irish University's in Dublin and the Scottish University's in Edinburgh. I seem to remember we beat the Irish and lost to the Scots. My second row partner was none other than Ray French from Leeds University who later played rugby union for England, subsequently turning to rugby league and representing Great Britain in many test matches. I met Ray again years later when he was the rugby master at Cowley Grammar School and I was on the touchline as a proud parent watching one of my sons playing against his team. By then he was the 'voice of rugby league' on TV and one of the game's great characters. We reminisced about an incident many years earlier, when in opposition, I had 'accidentally' stood on him when he was on the ground. I could see that he didn't think it accidental and knew that I was a marked man. Following his team's kick at a penalty goal I could see Ray out of the corner of my eye approaching at great speed, fist drawn back to deliver a knockout blow. At the last minute I ducked and he flew harmlessly over my back. The next second the final whistle went and we retired the best of friends to the clubhouse. If I hadn't seen him coming it might have been the end of a promising career.

Rugby in the sixties was organised very differently from today's professional game. Fixtures between clubs were based on historical rather than performance criteria. Club fixture secretaries were forever trying to persuade officials from the more prestigious clubs to grant them a game, and who you knew counted for much more than the first team track record. At the top end of the game, the way forward for aspiring players was via the County Championship and then onwards to a series of England Trials. The northern region included the six counties of Northumberland, Durham, Yorkshire in the east and across the Pennines, Cumbria (then Cumberland and Westmorland), Lancashire and Cheshire. Northumberland were especially fortunate in that they had their own county ground in Gosforth which doubled up as a greyhound stadium and meant that they had rather more financial resources to play with than most. None of this found its way to the players, but the 'committee men' seemed to enjoy themselves in style.

I was selected to play for Northumberland during the 1959/60 season and played in all five regional matches. The following year, I captained the side when Derek Morgan was injured and played a couple of games in 1962 just prior to getting married and moving down to London to start my first job. Again the results were no better than average and we didn't win the northern group in any of those three years. It was undoubtedly the highest standard which I played, and a tremendous privilege to be part of a team containing the best players in the area.

Rugby has been a central interest ever since my postgraduate days at Newcastle. One of the spin-offs was getting to know Prof Mac Cooper outside the formality of the Agricultural Faculty. He had, for some time, been President of the College Rugby Club and was also a county selector. His own rugby pedigree was impressive, having captained Oxford University and played for Scotland in the thirties. After returning to New Zealand after the war, he captained the Wellington Club to many famous victories and some of his fans maintain that he was unlucky not to become an All Black. Although he didn't have time for coaching the side he volunteered to turn out on Sunday mornings with the team to discuss the previous day's match (he was a keen spectator) and talk us through the finer points of the game. We got on extremely well and it was through his obtaining some tickets for a match between the Northern Counties and South Africa early in 1961 that I first met his eldest daughter, Barbara. She was a medical student at Newcastle and we sat next to each other during the game. In later years she was apt to remark that although she had heard of my rather wild exploits both on and off the field, she was surprised that I seemed more civilised than she would have expected!

Love and marriage

As a student and at school I had had a series of girlfriends, none of whom might be described as 'steady'. My priorities were firmly towards sport and nights out with the boys. When we needed partners for dances or balls, there were plenty of girls to choose from without the inconvenience of having to take any one of them out on a regular basis. On one occasion in early 1960, my current girlfriend was unavailable to attend the University Boat Club Ball at the Castle in Durham. One of the Prof's other daughters, Diana, was then working in the field laboratory at Cockle Park and despite my best persuasion wouldn't go either – she was engaged to someone in Edinburgh at the time. She did, however, suggest that I ask her elder sister, Barbara. Had she not met me originally at the South African match, I doubt very much that she would have agreed; however fate smiled and we had a memorable evening. The next Saturday happened to be another ball, this time in aid of a fund-raising effort for the rugby tour to Cornwall. I had already arranged to take Janice, my current girlfriend, but also persuaded a colleague at Cockle Park to take Barbara so that we might continue where we left off the previous weekend. It was not an action in retrospect that I was proud of, but it did herald the start of a lifelong partnership, which has never looked back.

As we became, in current terminology, an 'item', I was on the receiving end of quite a bit of friendly abuse. Some of this stemmed from my college friends, few of whom had steady girlfriends and saw my handholding in public a betrayal of our macho sporting image. Rather more leg-pulling emanated from the staff and fellow students at Cockle Park with regard to courting the Prof's daughter. As for the old man himself, I saw quite a lot of him with regard to the research work I was doing, as well as rugby events. When it became clear that Barbara and I were spending a lot of time in each other's company, he seemed genuinely pleased and encouraging, as indeed was his wife Hilary. From my point of view this was a great relief as had the vibes been discouraging, then it would have made life very difficult for both of us.

I became a frequent visitor to the Cooper home at Tritlington, a renovated farmhouse within a mile of Cockle Park. Mac and Hilary were generous hosts and immensely popular with all the staff both at the farm and in the local community. I also met Cynthia (Squint as she is known to family and friends), Diana's twin sister who subsequently married Dai Morris, my roommate in the Pele Tower at Cockle Park. Dai had graduated at Aberystwyth and had come to Newcastle to do a PhD in grassland management. Needless to say I got some of my own back when it became clear that he had designs on Squint, the remaining daughter.

In the summer of 1961 Diana married Martin Thompson, a farmer from Lincolnshire. This was the opportunity to meet most of the family, two of whom had come over from New Zealand for the wedding. Thelma and Madge were both elder sisters of Mac Cooper and with their respective husbands were enjoying an extended tour of the UK. On Hilary's side, as well as her two elder sisters I was introduced to Mops and Pops, Barbara's grandparents. They lived at Boars Hill, a very fashionable part of Oxford. The story goes that Hilary's mother did not approve of being called 'Granny' at the tender age of about forty, hence the sobriquet to smooth the waters. She was a formidable, rather intimidating woman, keenly interested to see that suitable young men escorted her granddaughters. I rather doubt I measured up to her aspirations.

Notwithstanding any shortfall I might have occasioned at this or other family events, Barbara and I were in love and with both our families' best wishes became engaged to be married at Christmas 1961.

During 1962 my research work was drawing to a close and as the threat of National Service had by then disappeared, my thoughts turned to serious job hunting. I didn't have any real ambition to get into practical farming, as without any widespread experience together with lack of capital, the prospects seemed remote. Neither did I relish the thought of a career in research or teaching, although I did go for an interview as a prospective manager of a small research farm owned by a feed merchant in Lincolnshire. Discussions with Prof revealed that the Milk Marketing Board might be a prospective employer as they were actively recruiting trainees for their new dairy farm costing and advisory service. I knew next to nothing about dairy cows but it seemed to offer an interesting opportunity and I duly applied and was delighted to be offered a job starting early in January 1963 at the then attractive salary of £900 per year. By this time Barbara had qualified as a doctor and was working in Newcastle General Hospital in a house job. My employment as a Trainee Consulting Officer would initially be based at the MMB head office in Thames Ditton about ten miles west of London. So we determined to go flat out, me to finish and submit my PhD thesis, get married shortly before

Christmas and move down to London after the holiday.

We were married at Hebron Parish Church near Cockle Park in December 1962 and spent our brief honeymoon in Eskdale in the Lake District. Some years earlier we had enjoyed a family summer holiday at The Woolpack, walking the fells, fishing in the gills and helping with the hay harvest on the attached farm. One of the owner's sons, Dick Armstrong, was a similar age to me and we had a great time rushing about. Amazingly Dick had turned up as an agricultural student in the same year as me and we became good friends over the three years at Kings College. He engineered a reduced rate for our honeymoon stay and somehow seemed to appear on all the photographs I took during the time we were there.

Agricultural politics

It is now necessary to spend some time discussing the agricultural industry in general and dairy farming in particular so that the reader might get some sense of understanding as to what I had let myself in for. The temptation to expound my own views on the political and controversial issues of the day is irresistible and I make no apology for so doing. You can always skip the pages or switch off.

Agriculture, once the biggest industry in the British Isles, has had a long and chequered history. Even within the twentieth century the viability of farming has swung from severe recession through the means of survival during two world wars, to a political seesaw of signals between expansion and contraction. Now any meaningful political control has been delegated to the bureaucrats in Brussels. Farmers, once revered for providing cheap, good quality food, often during extreme conditions, seem to have lost much of their respectability.

During the Sixties and early Seventies and in reply to Government encouragement to expand home production and save imports, UK farmers responded magnificently. The problem was that on joining the Common Market (now the European Union and referred to as the EU throughout), guaranteed prices rapidly produced unmanageable surpluses and the cost of these almost broke the bank. Taxpayers quickly came to the view that farmers were a protected species and ought to face up to the realities of life as had the shipbuilders and textile manufacturers, rather than rely on subsidies from Government either here or in Europe. But rather than bite the bullet, the euro-politicians introduced quotas in the early 1980s and UK farmers were constrained in their output, not determined by our own national demand but by the overall balance sheet within the EU. Since then, financial support to agriculture has changed direction yet again, away from subsidy based on price of product to a land payment system irrespective of the farming operation.

For someone who has to plan long term to manage a business based on growing cycles, animal production timescales and weather extremes, such continual moving of the goalposts is hardly conducive in helping to define a successful strategy. There are many, myself included, who wish that all subsidies to farming had been abolished long ago. This happened in New Zealand and although their conditions are very different, such a policy galvanized their farmers and their representatives to develop new markets and opportunities worldwide. Furthermore, it removed at a stroke the public criticism that the taxpayer was continually underwriting the industry. Like everybody else farmers had to stand on their own feet. New Zealand farming has never looked back. We, on the other hand, seem to be irrevocably immersed within the EU with its influential small farmer lobby and endless bureaucracy.

Milk Marketing Board

The MMB was established in 1933 following years of price chaos within the industry. As a perishable and also a bulky product, milk is an inconvenient as well as an expensive commodity to transport. By far the most valuable sector of the market was the fresh pint, delivered direct to the doorstep. In those days this accounted for the majority of the national production with the remainder being manufactured into cheese, butter, skimmed milk powder and an increasing number of added-value products. Prior to the establishment of the MMB, the dairy companies preferred to buy milk for their urban consumers as near to the point of sale as possible. These farmers, mostly in the Midlands and eastern parts of the country therefore flourished, with prices way in excess of their country cousins in the far west and Wales, who had no choice but to accept the price the cheese and butter manufacturer would pay them. In some cases they were unable to sell their milk at all.

The 1933 Agriculture Act established a statutory monopoly, which compelled all the dairy farmers in England and Wales to sell their milk to the MMB (milk producers in Scotland and Northern Ireland had their own Boards). This organisation then negotiated the best possible sale price for milk to the manufacturers and distributors within each sector of the market. The resultant average price for all the milk produced, irrespective of location, was then paid by the MMB directly to the producers at the end of each month. Although an oversimplification of the scheme, the outcome was an unqualified success as far as the majority of dairy farmers were concerned. At last they had some collective bargaining power through a board of democratically elected representatives who shared their interests. In the event of the seller and the buyer being unable to agree prices, then there was a formal process of arbitration. Just as important, farmers had a guaranteed market and a regular monthly milk cheque.

The Milk Marketing Board is now past history but in the 1960s it was an admired worldwide example of how a successful farmer co-operative could work. Most people unconnected with dairy farming thought the MMB was an offshoot of the Ministry of Agriculture. However, nothing could have been further from my mind than joining the massed ranks of the Civil Service. I might not know much about cows but I was well aware that the MMB was a dynamic and forward-thinking organisation and the bit that I was to become involved with was at the farm production rather than the marketing end of the business.

The Board itself consisted of fifteen farmer members elected on a regional or national basis by all milk producers in England and Wales. In addition, the Ministry of Agriculture appointed three members to allegedly bring in specialist expertise from the outside world. The eighteen-strong Board elected its own Chairman and Vice Chairman. It employed a Chief Executive and Executive Directors to run the vast operation both at Head Office and in the field.

Elected Board members had often made their name in National Farmers Union politics and there was usually keen interest when their three-year term was up for renewal. Most were practising dairy farmers or had handed over the running of their farms to the next generation and become immersed in local politics. Not surprisingly they were largely strong-minded and articulate. Their principal job was to secure the best possible milk prices for their constituents, and arguments with the trade representatives were frequent and often long-winded. But the organisation had a much wider brief than just milk price negotiations. They were responsible for milk transport, quality of product, paying producers each month, farm services' provision, and in running their own manufacturing creameries – later expanded and hived off to become Dairy Crest plc.

Trainee Farm Management Consultant

I started work for the MMB on 7 January 1963. January and February that year turned out to be two of the most severe winter months of the century. Married for less than a month we stayed initially with friends in Harrow before eventually managing to rent a furnished bungalow at Walton-on-Thames about five miles from the MMB Head Office in Thames Ditton. This was to be my location for an initial three months, prior to training in the field. The house was within a few hundred yards of the River Thames and built on brick pillars, presumably to avoid problems of damp. All the pipework supplying water and drainage were exposed to the elements under the floor and within a couple of days of taking up residence the freezing temperatures put paid to any running water. In fact I can recall the exact moment of the freeze-up as the outlet pipe from the bath blocked at about 10.0 p.m. By the morning the water left overnight in the bath was a solid lump of ice. At one stage we used the bath as a fridge to preserve perishable items of food. The day we left the house, six weeks later, I remember melting the iceberg with a kettle of boiling water and heaving it out through the bathroom window.

The bungalow had no central heating and coal for the fire was unobtainable. Our sole source of fuel was a car boot full of logs procured from Barbara's grandparents who lived in Oxford, and a few bags of coal from her uncle in Sevenoaks. Still we were newly married and spent plenty of time in the warmth of the comfortable double bed.

When I arrived at Thames Ditton as a complete novice in early 1963 there were over 100,000 milk producers in England and Wales. They milked over 3.2 million dairy cows and had an average herd size of just over thirty, although this figure masked a wide range from smallholdings with a few animals to large efficient units of over 100 cows. Average milk yield per cow was about 3,500 litres per year which added up to a vast volume of milk, all of which had to be collected from every farm and transported to the dairies 365 days of the year. It was a huge logistical problem made more challenging when snow and ice blocked the narrow lanes to many of these farms stuck out in the far flung corners of the country.

In 1963 the MMB had been in operation for thirty years. There were about 1,200 employees based at Head Office with probably twice that number working in the twelve Regional Offices or in the field directly with farmers. Organising the daily collection of milk and ensuring it went to the right dairy company was a vital and complex process. In addition there were scores of people working on the administrative and producer payment side of the organization. But it was not all bureaucracy, milk price negotiation and transport. The Board had decided some years previously that it should also help farmers improve their profitability by providing fee-paying services to enhance the efficiency of their businesses. The two main activities which by then were well established were milk recording individual cows and artificial insemination.

Although milk is a staple diet, it is surprising how little understanding there is about the way it is produced and about the animal which produces it. The modern dairy cow is the result of years of selective breeding, but it was not until the introduction of artificial insemination in the 1940s that the potential for increased milk production per cow really took off.

When I started work in the industry the average cow in England produced about 3,500 litres per year. By the time I left in 1996, the equivalent yield was over double this figure. In fact over this thirty-three year period the dairy farm business was transformed from a small, high labour-intensive, individual cow enterprise, to a large, highly mechanized business. Average herd size increased from about thirty to nearly eighty. Instead of being housed in individual stalls in cowsheds, the animals were kept in large groups bedded on straw or lay in sawdust-covered cubicles. Whereas the cows had been milked twice a day by a machine into a portable bucket and the milk transferred to churns, by the time I left the MMB, almost every dairy herd was milked (sometimes three times per day) through a sophisticated milking parlour. Not only did the parlour automatically record each cow's production, but it also dispensed supplementary concentrate feed on a strictly planned scientific basis. Furthermore the milk, instead of being manhandled in the churn to the end of the farm drive, was now directly piped from the parlour into a refrigerated bulk tank and the lorry arrived each day to pump it out and take it to the dairy.

In the winter when the herd was indoors for the best part of six months the staple diet had been surplus summer grass conserved as hay. Not only was this a laborious process both in making and feeding, the British weather usually saw to it that the product was inferior, especially in the high rainfall areas of the west. By the sixties most dairy farms had opted to make silage rather than hay, and by putting the conserved fresh grass in a large enough storage area, the cows could actually walk from their winter housing area and help themselves.

All these changes cost a great deal of money and they also had a number of longer term adverse side effects, which today contribute to the debate about animal welfare, pollution and food quality. But at the time it was Government policy to encourage farmers to expand production, not only in milk but in everything else. These were the days before Britain's entry into the EU and its Common Agriculture Policy. We imported massive amounts of food and the reasonable assumption was that by producing more of it ourselves we could save substantial import costs. To this end the Government introduced all sorts of financial incentives for expansion. Grants and subsidies were available for machinery purchase, fertilizers, drainage and fencing. Special tax allowances were introduced to offset capital expenditure and even the cost of farm business recording was underwritten by Government grant. The message was clear – modernize and take advantage of the new technology. The opportunity for organizations such as the MMB and the Ministry of Agriculture Advisory Service to help farmers implement this policy was self-evident. I had had the great good fortune to join the Breeding and Production Division, of the MMB at just the right time.

The MMB Head Office was an impressive three-storey building with a colonnaded entrance, spacious grounds and a white-collar workforce well removed from the business of farming. The Breeding and Production Division, of which I was a very junior part, occupied a large ground floor office. I was allocated a desk and given various fairly mundane tasks, either adding up columns of figures or reading about the great organization, which I had recently joined. No one seemed terribly interested in what I was doing so I set about identifying key people and asking them if they could spare half an hour or so to explain what they did. Most people love talking about themselves so within a short time I learned a lot about the operations as well as the internal politics of the institution. I had to endure six weeks of this so-called induction before being allowed out in the field as a trainee with the Consulting Officers during their daily round of farm visits.

The Consulting Officer Service was a fairly new innovation. The concept originated in New Zealand and was introduced by Alan Stewart, himself a Kiwi who, after completing his studies at Oxford, joined the MMB as an Extension Officer. An organization in New Zealand, similar to the Board's Breeding and Production Division employed a number of agricultural graduates to visit dairy farmers to help operate their cattle breeding and advisory services and Alan's view was that the MMB should provide a similar pool of talent. This group was established in the early sixties and the twenty or so Consulting Officers were scattered about the country with their initial main task to help locate top quality dairy cows, which could be mated under contract, to produce potential young bulls to be used within the Artificial Insemination Service.

Cattle breeding programmes

Although the MMB had pioneered artificial insemination for dairy cows there continued to be vigorous debate as to the source of the bulls which should be used for the national stud. In fact this whole area is a good example of a conflict of interest within the so-called spirit of farmer co-operation. In order to comprehend, I'm afraid it is necessary to explain the basic concepts of a cattle-breeding programme. Again for those with no interest in the subject they should fast forward a few paragraphs.

An essential requirement of a successful dairy herd is to ensure that the cows within it have a calf every year. It is amazing how many people unconnected with farming still think that a cow only ever has one calf and produces milk continuously for years. Without the physiological stimulus of birth, milk production would gradually tail off and the whole enterprise would quickly become hopelessly uneconomic. In practice about eighty days after calving the animal is either served naturally or by artificially insemination. Following a nine-month gestation period she produces another calf and the cycle is repeated for as long as possible. The modern dairy cow has or ought to have a well-earned rest for about two months before she calves so that she can build up body reserves ready for another lactation. Although some cows produce milk for many years, the average is only about four or five, as many have to be culled due to disease, infertility or low production. It therefore follows that in order to maintain numbers, the farmer has to have about 20% of new milking animals or heifers coming into his herd each year. Some buy these replacements from the market but most prefer to rear female calves bred from their own cows. So assuming the farmer wants to improve the genetic potential of his herd to produce higher yields, the whole process of selecting the right bull as father to succeeding generations is of critical importance. It continues to be a subject of unending argument and opinion between dairy farmers.

Until the introduction of artificial insemination in the early 1940s the farmer had to buy or borrow a bull for natural service. He had no way of knowing whether this animal would help to improve the capacity for milk production in his herd or not. The seller of the bull on the other hand would provide impressive records of milk production from his own herd as well as comprehensive pedigree lists of the aforementioned bull's ancestors, thus justifying the highest price possible for the purchase. The fly in the ointment was the comparative ease with which an unscrupulous breeder could increase individual milk yields and therefore official records by feeding massive amounts of concentrated dairy rations to his cows. The sales pitch was that bulls from his high yielding cows were of superior genetic merit, although in fact the level of feeding (which was hopelessly uneconomic) was the main reason for their level of production. In effect it was a con trick and many of the resultant progeny emanating from expensive bulls failed miserably to live up to expectations.

The introduction of artificial insemination in the 1940s was not a universally popular concept. There was vigorous opposition, principally from church leaders, as to such an unnatural practice – even in cows. Despite, however, the raised voices of the bishops in the House of Lords, common sense prevailed and the MMB began to develop both the technique and the infrastructure necessary to provide an 'on farm service' by trained staff. With the advent of frozen semen a few years later, the scope for exploiting the genetic potential of the best dairy bulls in the country opened up a tremendous opportunity to improve the national dairy herd.

It was one thing to develop the means to artificially inseminate, but quite another to select young bulls which would actually provide the genes to improve their daughters' chances of producing more milk than their mothers. What was needed was a scientifically designed and wholly objective method to identify young bulls, based on their true genetic merit for transmitting increased milk yield to their female calves. As might be expected, many of the less scrupulous pedigree breeders saw this as an unwelcome challenge to their traditional market place.

Eventually a Progeny Testing Scheme was launched with the MMB providing the scientific evaluation as well as the substantial funding required. The scheme required the identification of young pedigree bulls from which semen was used in co-operating herds to produce daughters, which when they began to produce milk, could be individually milk recorded. These records were then compared with daughters from other bulls used in the same herd, thus removing the effect of feeding bias. The daughters from these promising young bulls were also evaluated for physical attributes such as udder attachment, healthy feet and even temperament. Provided they passed all the Ministry health tests the best bulls could then enter the national stud and their semen used to father many thousands of daughters in dairy herds throughout the country. The programme was a massive one and involved housing and rearing the young bulls under test until their daughters had themselves produced a full lactation. It could take as long as six or seven years from identifying a young bull until his daughters could themselves be milk recorded to see if they were significantly better than their contemporaries. Only a handful of the 100 or so bulls reared each year were eventually selected, the rest were simply slaughtered. The total costs of the scheme ran into millions of pounds per year.

As the MMB was a farmer co-operative, with a number of prominent pedigree breeders on the Board itself, the staff of the Breeding and Production Division had the unenviable task of reconciling self-interest on the one hand and commercial reality on the other. Each AI Centre had a farmer committee as well as a panel of breeders who inspected those daughters of bulls, which had promising results. The Consulting Officers were at the sharp end in as much as they organised the farm inspection visits and made recommendations with regard to whether young bulls were to be used or not. Clearly the prospective financial gain for the owner of a top performing young bull was considerable as the MMB paid out handsome amounts of money to buy these young bull calves to put into the testing scheme. It was not unknown for individual farmers on the panels to gang together to either veto a decision on an animal from what they considered to be 'an undesirable source' or conversely push one of their own breeding. It was a political minefield and one I, still wet behind the ears, was anxious to avoid.

The Low Cost Production Service

I had joined the MMB because a new costing and advisory service had recently been launched to provide farmers with information to manage their herds more efficiently. It was to be called the Low Cost Production Service or LCP to give it its obvious acronym. The Consulting Officers, all agricultural graduates, were used to help set up the service in 1962. The plan was to re-organise the field staff into Livestock Officers who would deal with all the breeding work, whilst the Consulting Officers would run the costing scheme and provide appropriate management advice.

My first taste of what would be my job for the foreseeable future was to be two weeks in East Anglia. I didn't know then that twenty-five years later I would be plunged headlong into the breeding business to face the ever escalating debate on why the UK had fallen so far behind the rest of the world in producing bulls of superior genetic merit.

The involvement in AI and now a dairy herd costing and advisory service within a national co-operative of over 100,000 dairy farmers, primarily set up to market milk, might appear to be stepping outside the MMB's terms of reference. The Board, however, took the view that it should provide additional help to its members to improve the efficiency and profitability with which they managed their enterprises as well as to secure for them the best possible milk price. The one condition which was sacrosanct within this philosophy was that the user paid the cost of the service. It must not be subsidised by the non-user. This was a fundamental criterion and meant that the AI Service and now the LCP Service had to be run commercially as separate businesses. I welcomed this. It seemed to me to be the best way to operate and it was the main reason why I had sought employment with the MMB. Unless the client was prepared to pay, we weren't good enough and we would be out of a job.

The new LCP Service, the forerunner of what is now a well-known international farm management consultancy company, consisted originally of a pre-planned monthly visit to the farmer member. The aim was to collect some simple milk production data and to calculate a few financial margins based on milk output and feed input. These numbers were then circulated in the form of a coded league table for the area and the client could estimate how he was doing compared to his neighbour. The visit invariably included a walk round the dairy herd, the farm and the buildings in order to discuss such matters as breeding policy, feeding regimen and all things related to the business.

The key to success was to persuade the farmer that you knew what you were talking about. This took a bit of doing as most of us were in our twenties; we arrived, fresh-faced, in a company car with folders and slide rules intent on persuading our clients who might have been doing their job for much longer than we had been born. And on top of this they had to pay for the privilege.

When I look back it does seem incredible that we survived those early years. We were learning as we went along. The essential skill was to listen to the client. He joined the scheme for all sorts of reasons. He may have been in desperate financial trouble and didn't know where to turn. He might not have had the courage to tell his wife or family that they might go bankrupt. He may be successful and wanting to expand and needed a business plan to put to the bank. He may have a small farm tenanted with only twenty cows or he might own a country estate with a herd of over a hundred together with beef cattle, pigs and vast acreages of cereals. Or he just wanted to show off how good a farmer he thought he was. I say 'he' but it could just as well have been 'she'. Many wives kept the books and were closely involved in the physical work on the farm. There were also a number of female farmers in their own right. Whatever the circumstances each situation was different and the first job was to find out the real reason you were there.

Farming is a lonely job. You often work long hours on your own with no one to share your ups and downs. After a few years doing my job it came as no great surprise to learn that, in the extreme, farmers have the highest suicide rate within the population.

Once you had an idea of the problem, the next step was to try and establish some sort of credibility. They all knew that you were a greenhorn and a figures man not a practical farmer. You had to try and convince them that you did know a bit about dairy farming. Walking round the farm was a golden opportunity to do this – not by offering advice off the cuff but asking leading questions and offering informed comment. You had to build confidence so that they felt you really might have something to offer. Above all, you had to establish some sort of rapport so that when you next rang to make an appointment you were welcomed rather than rejected. I had spent a couple of months at Head Office and I couldn't wait to get away.

Alf Francis was one of the more experienced COs and had piloted the new service in his patch of the country, which was virtually the whole of East Anglia. Alf was to be my first mentor. He was a small, rather emaciated man, with a wonderfully dry sense of humour. It was clear that he was well respected in the area having originally been involved with the bull-breeding programme. We covered vast distances between the farmer members and what struck me most forcibly was the enormous range of enterprise we visited. One day we spent an hour with a man who milked about twenty-five cows doing all the work himself. Later in the morning we visited an estate belonging to one of the landed gentry whose wife, Lady Someone or Other, kept Jersey cows. She wasn't too bothered whether they made any money as long as they looked nice in the park. In the afternoon we went to a boys' approved school where they had a farm and a dairy herd. The boys did most of the work, which rather complicated the labour cost calculations.

My fortnight with Lindsay Patterson in Devon provided a sharp contrast. East Anglian dairy herds are often a subsidiary enterprise within a mixed arable farming system. In the far west, the dairy herd was usually the sole source of income. Also they had to deal with the wet weather, especially as they relied then mainly on hay-making in the summer months. Lindsay was a Wye College graduate and knew Mac Cooper when he was there as Professor of Agriculture, and gave the appearance of a rather well bred young man with a posh accent. This was OK for the 'toffs' but a bit off-putting for the Devonshire yokel. He had a laid-back approach to the job and whilst unquestionably able, preferred not to take life too seriously. He was the most remarkable mimic and storyteller I've ever met so he had me rolling about on the floor on many occasions.

By now it was approaching spring and my next assignment was to be a six-month training period based in Gloucestershire with another Consulting Officer called David Roberts. David lived in Cheltenham and Barbara and I decided to rent accommodation in the same town and perhaps get some furniture of our own. We located a ground floor flat in Eldorado Road with one large sitting room, two bedrooms, a bathroom and a kitchen plus shared use of a large garden. Barbara was expecting a baby and a routine examination revealed that her blood pressure was far too high. This meant hospitalisation and the start of what was to be a most traumatic period of our young lives. It was decided that the baby must be induced and on 19 May our daughter was born in Cheltenham General Hospital. At just over 4lbs, the resident paediatrician told me that she was not expected to live and advised that she should be christened in hospital. We were distraught and had to decide on a name immediately. We had had discussions some weeks previously and agreed on Susan but additional alternatives had been forgotten in the events of the previous few weeks. Thus a local clergyman was called to the Intensive Care Unit where he officiated with a nurse and myself in attendance. Barbara wasn't well enough to participate. Thankfully the paediatrician was wrong. She hadn't reckoned with Sue's fighting spirit. It was an emotional ten days but with expert nursing care she pulled through and suddenly we were a family.

By the autumn, the powers that be had decided that I was ready to take over an area and I was promoted to full Consulting Officer status and allocated a company car. I was to take over Cheshire as well as a few customers in Staffordshire and Shropshire. We were delighted to move to the north-west, right in the heart of dairy farming country and we set about trying to find somewhere to live.

CHAPTER FOUR

The Rest of the Sixties

A FTER A NUMBER OF RUSHED WEEKEND TRIPS from
Cheltenham to Cheshire, Barbara and I managed to purchase a small,
sandstone-built farmhouse in Manley, about ten miles due east of Chester.
Originally a smallholding, the owner, a local nurseryman, retained the land
and sold us the house, adjoining buildings and orchard for the heady sum of
£3,900. We moved with few belongings but with great excitement to Bay Tree
Farm in October 1963.

My first clients

My predecessor in the area had been Peter Woodriffe. He was an out and
out livestock man and couldn't wait to hand over his LCP clients to me.
They were mostly in Cheshire, but a few farmed in the surrounding counties
of Staffordshire and Shropshire. I had a Hillman Hunter, a smattering of
knowledge and about thirty customers to whom I was expected to provide a
professional service. It was an awesome task, but youthful enthusiasm knew
no bounds and I was soon ringing them up to propose farm visits. Consulting
Officers worked from home which seemed to be a significant advantage at
first, but in practice meant that you never really left the job behind. Arriving
home in the evening meant going through the day's mail and as I got to know
my customers, the telephone used to interrupt meals and weekends on a
regular basis. Farmers appreciated the fact that we were always on call even
if our wives didn't. Compared to nowadays, the luxury of being able to speak
to someone rather than a disembodied voice telling you to select a series of
options or a well-meaning but unintelligible respondent from Mumbai, meant
the telephone was the vital means of communication. Thank goodness we
didn't have to deal with email.

The first priority was to get to know my clients. The information we
collected (there were no calculators, everything was based on mental
arithmetic) was very basic and often focussed on the subject of feeding the
most economic ration to the herd. Again some reference is probably necessary

to assist the non-agricultural reader understand the basics of feed rationing and the economics behind it. Fast forward again.

Dairy cow rationing

The cow is a ruminant; that is to say it has a series of separate stomachs through which grass, hay, silage and other seemingly indigestible feeds pass on the way to the intestine. The first stage in the process is for the cow to regurgitate her meal and chew it a second time to aid exposure to the bacteria lurking in the subsequent stomachs. Very much like a fermenting vat, the food is then broken down into sugars, fats and proteins which are absorbed through the gut and enter the bloodstream. The physiological design of the cow is to transfer many of these nutrients, via the blood, to the udder to produce the milk – the remainder supplying the daily living needs of the animal. This process is the same with all female ruminants such as sheep, goats and many undomesticated species in the wild. The milk comes on stream when the animal produces its young and gradually tails off as the offspring grow up.

The modern dairy cow has been bred over generations to produce massively more milk than her calf could ever drink. Whilst some might argue that she has been exploited, the farmer must try and strike the right balance utilising the resources he has available on the farm. He might aim for differing production levels but he will also need to ensure that his animals, apart from being well fed, are comfortable and healthy if they are to achieve a long herd life. In the end it boils down to economics and it was one of our jobs to help the farmer find that balance on his farm.

In simple terms the cow's daily feed ration will consist of grazed grass in the summer and conserved grass, usually silage, in the winter. There are countless other feeds that can be added or substituted depending on their availability, but in the western parts of Britain, grass is the staple diet. It may come as a surprise to learn that not all grass or silage is of the same nutritional value. As the growing season progresses through spring and summer, the grass, if left to grow, becomes more fibrous and less digestible. By regularly eating or cutting it during the season the skilful farmer can ensure that the quality of the re-growth remains high and his cows produce milk at a lower cost, thus helping to maintain profitability.

The final piece of the jigsaw in cow feeding is to provide the right quantity of supplementary feed. The object here is to replace forage with concentrates. Unless this happens the cow will fill herself up with bulky wet feed which will be insufficient to meet her nutritional needs for a high milk yield. Because the concentrates are much more expensive, it follows that they must be carefully rationed to each animal dependent on the amount of milk she happens to be producing. It was often the case that farmers in their enthusiasm to raise milk yields fed their cows far too many concentrates so that the extra feed costs exceeded the value of the additional milk produced. These were the days of cattle feed sales representatives doing their rounds and holding farm visits to demonstrate the virtues of their particular product. It was both a complex and a confusing subject and always produced vigorous debate. One of our main jobs was to help the farmer disentangle the facts from the fiction so that he, rather than the feed manufacturer increased profit. Naturally enough we were considered aliens by many of these commercial companies, some of whose senior staff would write to Board members and complain that we were spreading misinformation.

The job

So the MMB Consulting Officer's job, having established a dialogue with the client, was to visit him on his farm on a regular monthly basis, record information about his dairy herd and work out some simple financial indicators as to whether he was making any money on that particular day. This exercise opened up countless avenues up which one might travel to discuss his grazing system, silage-making techniques, bull selection choices, mastitis problems and countless more. The fact that we were visiting two or three farms a day, attending farm walks and perhaps discussion group meetings in the evenings helped immensely to increase our own knowledge. You could soon spot an opportunity or identify a mistake you had seen on someone else's farm which could be applied with benefit to the case in hand. Unlike competitive commercial business, farmers were more than willing to show you their successes and even their failures. In a sense we didn't have a stock of answers, rather we were simply passing on other people's experience.

We also recorded details of other so-called variable costs of milk production, as we went round farms. The object was to produce a financial summary of the herd performance at the end of the year. Total sales of milk, calves and cull cows provided the income from which the feed, fertiliser and other direct costs were deducted to produce a 'gross margin' for the dairy herd. It was not meant to be a calculation of profitability as all the overhead costs of labour, machinery, rents and financial charges were ignored – as indeed were the contributions other enterprises might make towards the business bottom line. The theory behind the gross margin analysis was the assumption that if you could find ways and means of increasing it, the overhead costs would stay broadly the same and the client would be making a higher overall farm profit. It may be, of course, that he was an expert dairy farmer and his pig enterprise was losing hand over fist. So whilst you might eulogise with him about his wonderful dairy herd, he couldn't understand why his bank account continued in a downward direction.

At the year-end visit, the challenge was to formulate a budget for the succeeding year and as a result suggest longer term changes which might lead to an improving financial result – a strategic, rather than a tactical, approach. This might include expanding the number of cows in the herd, investing in a more efficient milking system, conserving surplus grass as silage rather than hay or a combination of a number of these options. It didn't, however, include the key question as to what to do with the pigs in the above example or whether he might be able to afford a new tractor because his current one was always breaking down.

As the years progressed it became increasingly obvious that budgeting on the basis of insufficient information was fraught with danger. We didn't know the degree of indebtedness the client might have with the bank or mortgage companies. We could only hazard a guess at the contribution other enterprises made to the whole farm, whether they be pigs, hens, cereals or anything else. To offer comprehensive advice on the way forward, without detailed analysis of the total business profit and loss account, balance sheet and cash allocation of funds was therefore exceedingly dangerous and could well land us in very deep water indeed. The LCP Service had to move into full farm accounting, not for tax purposes, but to provide information upon which to base reliable and professional advice.

This initiative rested with the boffins at Head Office and all manner of trial recording systems were tried out in the field. We already had some farm recording staff who visited the farms to collect data, but this move to fully accountable and reconciled farm records meant a much greater degree of training and understanding. We did try farmer input records but it just wasn't possible to get them to conform to disciplined systems of recording.

So by the end of the second part of the decade we had a Total Farm Business package up and running. Clients had the choice whether to stay on the dairy herd option only with its obvious limitations, or pay considerably more for a fully comprehensive service. Fortunately this move coincided with the Ministry of Agriculture offering farmers a grant which covered most of the recording cost, although the consultancy input was a separate charge.

Looking back at the names of some of those early clients nearly fifty years on, some are now no longer with us, others are still good friends, and most of them, like me, have now retired and their sons or daughters run the farm.

On the lighter side

Being an MMB CO was a wonderful opportunity to see farming and indeed life at the sharp end. It was hard work, hectic and hugely enjoyable if not at times hair-raising. A colleague for many years, Alwyn Hardy, a CO in Wiltshire, wrote a little handbook in the 1980s called *Marginal Notes*. Alwyn was a jolly, rather well spoken man, with a wicked sense of humour. He also had a healthy disregard for all the MMB administrators who seemed to plague his life. He simply got on with the job and his farmers and ignored everyone else. *Marginal Notes* charts twenty-five years of farm management consultancy under the auspices of the MMB. Now well into his eighties, Alwyn still delights in reminiscing about the old days and the many stories which befell our colleagues as they juggled their job round their clients and the MMB regional managers who were supposed to be the senior administrators in the region. They were rather jealous of these young bloods rushing about doling out advice on farms, whilst they were stuck behind their desks making sure the well-oiled MMB machine ticked along.

Concerning the difficulties of reliable and accurate farm recording, Alwyn recounts a few typical examples of the day:

Worst of all were the straight swaps – a litter of piglets for two old bicycles, a lawn mower and six dozen eggs. There were also the undisclosed sales kept hidden from the taxman. I once spent a long time trying to reconcile barley sales with the harvest tonnage and, try as I might, I always finished up with five tons unaccounted for. When, finally, in desperation I admitted to the farmer that I could not account for this barley, he remarked quite calmly 'You're sitting on it'. He had used the proceeds to buy a new carpet and had simply kept quiet about it. Clients were not altogether convinced that these records would not somehow end up with the Revenue – but they never did.

He goes on to relate:

Looking at the Consulting Officer Conference photographs reminds me of the only CO who ever, as far as I know, actually suffered physical assault. Working for the Breeding and Production Division is not generally considered particularly hazardous. Savage dogs and bad tempered bulls may be encountered from time to time; there used to be the distinct possibility of contracting brucellosis from the copious and well meant cups of tea offered in some farm kitchens and doubtless the increasing number of ladies now in the service experienced their own problems. But on the whole it was not a dangerous occupation. It is true that there used to be a member buried in the vastness of the Brendon Hills who frequently relieved his frustrations by throwing plates at his wife, some of which sailed through the kitchen window, which could make approaching or leaving the farm a bit exciting. This was incidental, no personal animosity was intended. But the previously mentioned individual on a farm visit once so enraged his client that he was picked up by the scruff of the neck and the seat of his trousers and literally thrown out of the door, with a warning that, if he ever came onto the farm again, he would be met with a shot gun.

I remember a couple of examples of surprising incidents from my own time on farms in those days. One was a mid-winter visit to a small farm near Audlem in Staffordshire. The client was a rather unpleasant man, forever complaining about how unfair life was and that everyone from the MMB who came to his farm to check up on his hygiene status or test the milk to see whether water had been added, was bent and out to get him. I dreaded going there and could never understand why he bothered to use the Service as he took not a blind bit of notice, and the farm and animals were in a sorry mess. During our session in a draughty kitchen there was a loud knock on the front door. The farmer's wife came into the kitchen and told Denis that there were a couple of people who wanted to see him. I could hear raised voices and feared that my visit was going to terminate rather quickly. In fact he was arrested by two police officers, handcuffed and charged with stealing geese from a neighbouring farm. They promptly marched him off to the Black Maria and I never saw him again!

The other story concerns a small farmer up in the Staffordshire hills near Biddulph Moor, way out in the wilds. This was unforgiving country and a hard slog to make a return on the back-breaking work involved in a one-man band doing all the milking, calf rearing, feeding and field work. The cows were housed in a traditional byre and most of their food had to be carried to them individually. He never had a holiday and I doubt he had strayed more than five miles from his farm for the last twenty years. Furthermore he was getting on and his wife was infirm. They had no children. I had prepared his annual summary and had made an appointment to visit him on a mid-winter afternoon. His wife had lit a fire in the small front room and it was there we repaired to discuss the figures and work out a budget for the next twelve months. With my head in the papers I was explaining some, no doubt unfathomable financial calculation, when I looked up to elicit his opinion as to my proposals. With horror, I found that he had lapsed into a deep sleep. No amount of coughing could wake him up and much to my shame I slunk away to leave him with his dreams of what might have been had fate dealt him a fairer hand.

An essential priority for the MMB was to ensure that a total of about 25 million litres of milk was collected every day throughout England and Wales and transported to the processing dairy. This was a massive logistical task and co-ordinated by the administrative staff in the Regional Offices spread across the country. About half of this vast volume was collected by specialist contractors with their own vehicles but the rest was picked up by the MMB's own transport fleet. Being in the transport business ensured that contracts with independent hauliers were keenly priced. Occasionally there were staff problems and even strikes as MMB drivers tried to exact improved wages and conditions. This necessitated other staff being drafted in to drive lorries and navigate the routes to the farms. Consulting Officers were often first in line for duty. This was before the days of HGV licences and the only training available was a 200-yard stretch in the car park before emerging with an eight-wheeler onto Preston High Street. There was no power steering, the roads were often very narrow up in the hills and we were frequently collecting milk from isolated farms in the middle of the night. Inevitably gateposts and walls took a heavy pounding, but the farmers were delighted that we got through as their tanks were overflowing and the morning's milk had to go somewhere. I learned one lesson quickly – never pull onto the grass verge to let someone pass. I ended up having to get my load pumped out when I gave way to someone in a hurry driving a Jag.

Perhaps the most famous story was that of a colleague driving up the M6 with motorists making strange signs as they passed him. He assumed these to be acknowledgements by the locals for his commitment to the job in hand. When he got to his next pick up point it became clear that the connecting hose from the lorry to the bulk milk tank had come loose and had been weaving in his wake across two lanes of the M6. In another incident, the hapless driver forgot to disconnect the hose from the empty bulk tank before driving off down the farm lane with it bouncing behind him.

The commercial aspects of the Service

One vital aspect of my consultancy job was to retain a full workload. To meet costs and produce a modest profit we needed to sell about 200 days of time on farms each year. This might not sound a high figure, but by the time holidays and weekends were deducted it left only about thirty days to do everything else. Quite a lot of office work was necessary in preparing budgets and writing letters to back up your deliberations on the farm. I found I got back home after a day of farm visits and then, on average, worked a couple of hours in the office most nights and sometimes at weekends.

It didn't really matter how many clients you had, although the average was about sixty, each having just over three days of time throughout the year. Sometimes these days were split down into a regular two-hour meeting on the farm which worked out at a short visit each month. Other clients preferred half-day visits lasting for a whole morning or afternoon. There was quite a high turnover of clients for a variety of reasons. One of the main ones was staff changes. Clients generally disliked change, especially if the CO leaving had done a good job. It also has to be said that some farmers thought the fees were too high or the CO was simply not good value for money. Whatever the reason there was a constant need to recruit new clients to maintain workloads. Unsurprisingly some COs had virtually no resignations. They had been there a long time and had built up an impressive reputation in their area. Prospective clients would contact them when they had a vacancy and would not accept visits from other colleagues. There were others who lost nearly half their clients during the year. Usually they had to go – unfair dismissal and tribunals were unheard of. It was a hard commercial world.

I was gaining confidence but always on the lookout for new business. One thing I knew I needed to do was become better known, especially in Cheshire. I needed to organise farmer meetings and farm walks. I needed to speak at events, write articles in the local press and be noticed. Public speaking was a terrifying prospect. You could get away with pontificating on a one to one basis, but in an audience of farmers you had to be on the ball. I resolved to join the local Grassland Society and Farm Management Association and at least try and ask an intelligent question at each of their meetings. Soon I began to organise discussion group meetings for my clients.

They weren't grand affairs. We usually met in a pub and people would bring their figures and share them with the rest of the group. There would be about twenty participants and sometimes I persuaded someone with specialist knowledge to come and give a talk followed by lively discussion. On one occasion I invited an academic from Nottingham University to come and talk about herd fertility. There were about thirty there that night and the landlord had lit a roaring fire in the lounge bar. There was always a faint smell of cow slurry or silage when you got a crowd of dairy farmers together in a confined place and this night was no exception. The Chairman for the evening was to introduce the speaker and went on for much too long extolling the importance of the subject and the research work about which the speaker was to tell us. I was sitting at the top table and I began to notice, with some alarm, that quite a number of the audience were beginning to nod off. One rather elderly gentleman in the front row even began to snore. Sitting next to him, a lady who I assumed to be his wife, suddenly gave him gave him a sharp dig in the ribs. Whereupon, the man struggled to his feet, produced some crumpled notes from his trouser pocket and proceeded to thank the speaker for such a fascinating talk. That certainly woke up the meeting and it turned out to be a great evening, if not a memorable one.

Occasionally we held meetings on a member's farm. We would examine his cows, walk over his fields and discuss the ways he planned to make improvements and increase profits. These were of tremendous value as everyone actually saw for themselves what might be possible and could then relate that back to their own circumstances. They learned from each other – I was just the catalyst.

Spy Hill Farm

Early in 1965 I saw a sale notice at the end of a farm drive about a couple of miles from where we lived in Manley. The thought of eventually getting a farm had crossed my mind as a number of my colleagues ran small enterprises in addition to their consultancy job with the MMB. Provided that the day job came first and you had a full quota of clients, then this activity appeared to be acceptable. In fact it added an extra bit of credibility as clients saw you struggling with bad weather and soggy fields just as they had to. Such dreams, however, were for the future – we had very little money at the time.

Nevertheless I went to have a look at the place with a newly acquired farmer friend, more out of curiosity than real ambition. The land totalled about seventy-four acres with a farmhouse, a traditional cowshed and some outbuildings. The owner, a widower, was well into his seventies and the overall impression was one of dereliction and decay. The house was almost uninhabitable. Part of it was used to store potatoes and apart from doors hanging off hinges, the place hadn't seen a lick of paint for years. The so-called garden was littered with decomposing rubbish and old ramshackle hen houses. The cowshed was in reasonable condition and the owner's son who had been milking the cows, no doubt hoped to take over the farm when his father either retired or died. It seemed, however, that they had fallen out and that was why it was on the market.

The land was untypical for Cheshire in that it was anything but flat. The subsoil was pure sand with clay caps on the top of the banks. In earlier times it had been the practice to dig out the clay or marl and spread it on the lighter, sandier land to improve soil structure. Thus there were a number of marl pits on the farm, mostly filled with water. The owner had, until recently, been a tenant of the Ashton Hayes Estate, and they had negotiated a contract for sand removal direct from about fifteen acres of the farm. The extracting licence had expired and the land was being re-instated at the time of our visit. Having purchased the farm from the Estate, it was clearly the farmer's intention to apply for a new sand extraction licence on fourteen acres of land he planned to retain and which was situated directly next to the farmhouse. So this was a significant downside risk to any prospective purchaser.

I came away from the farm with two main conclusions. First, it was in such a bad state of repair, it might sell cheaply. Second, the location on the very edge of the beautiful Delamere Forest was quite outstanding. Had this opportunity to remain a dream because the timing wasn't right or should we take the risk and at least have a go for it?

Barbara was the one with the real vision. By then we had two toddlers; we had only been at Bay Tree for just over a year and we were renovating the house. Also I had joined Chester Rugby Club so my Saturdays in the season were likely to be taken over by that commitment. It says a lot about her foresight that she was enthusiastic; had she not been I'm sure we would have let the chance go by. Who knows there may not have been another opportunity as land prices were destined to rise steeply over the next few years.

I did the sums, talked to my father and brother about loans. We had the value of the house we lived in and in the days when you could still talk to a local bank manager, I persuaded him to let us have a sizeable overdraft. Together we set a maximum bid price and so on 23 March 1965, David Dodd, my farmer friend and I went to the auction at the Bowling Green Hotel in Chester.

The auctioneer started the proceedings by inviting a bid of something over £20,000 – way outside my reach and I thought that was that. There was shuffling of feet but no takers. He progressively came down until he got to £10,000 when someone put up his hand. The bidding rose steadily but slowed at £13,000. We were now in the realms of possibility but it became a game of nerves as the process stalled at £14,500. I held my breath and signalled £15,000, the maximum number we had agreed. Silence. The auctioneer, try as he might, simply couldn't move the price higher and he knocked down the sale to 'that young man in the back row'. It was a life-changing moment. As I went up to sign the cheque for the £1,500 deposit I was seized by alternating waves of elation and panic. I had just bought a farm, how the hell was I going to run it and do my job as well?

Contract rearing dairy heifers

I had to find a system which required virtually no working capital. All our money was tied up in the farm purchase. The main enterprise was to be rearing dairy herd replacements. At that time dairy farmers were beginning to intensify their businesses. Most of my clients wanted to rear the female calves from their dairy herds, having used semen on their cows from proven AI bulls with high genetic potential to improve milk yields. The alternative meant they had to buy replacements from the market or from a cattle dealer. Too often this resulted in introducing disease into the herd and most thought the risk was too high. Besides, cattle breeding is both an art and a science. In the pedigree world, breeders spend their whole lives thinking, looking and working with their animals. It is their hobby as well as their occupation. Pedigree or not there is a great sense of achievement in improving the appearance and productivity of the herd over the succeeding generations. This is one of the reasons why, in the event of say, a foot and mouth epidemic when cattle have to be slaughtered, the loss of a dairy herd which has taken years to build up is such a tragedy for the farmer and his family.

The rearing of dairy herd replacements from calves to the time when they can themselves produce milk takes about thirty months. This means that some of the land has to be devoted to this task instead of perhaps expanding the number of milking cows in the herd and thus potentially adding significantly to turnover and profit. Thus there was an opportunity for these herd replacements or heifers to be reared off the farm by someone else. They would remain the property of the farmer and he would pay a monthly fee to cover the feeding and other associated costs until they were ready to return to the original herd to begin their milking lives. For me this meant no capital was needed to buy stock and the monthly contract fee financed the purchase of seeds and fertilizers to grow grass and corn. This was the enterprise I had in mind.

From local farm sales I bought some second-hand machinery – one ancient tractor, some field harrows and a flat roller. All the rest of the field work was undertaken either by my pal David Dodd, who had a dairy herd within a few miles, or by agricultural contractors. I had rearing contracts with three clients and by the summer we had about fifty heifers on the farm. We had sold our house and made a profit of over £1,000 so we began to feel a bit more confident, although my father in later years was apt to remind me that he had said goodbye to the money he lent us.

The practical problems were many and as I was out doing my job every day, there were bound to be crises which Barbara had to resolve. Most of these were to do with our animals getting out, either onto a neighbour's land or worse onto the roads. The fences we inherited were virtually useless and all needed replacing, which meant not only money but also time, both of which were in short supply.

Many tales of family folklore surround stories involving Barbara, often with a child in arms followed by up to three others, observed rounding up stray animals in the neighbouring parish.

One of my responsibilities was to get the older heifers served when they were about twenty months old so that they would be ready to produce their first calf and return to their owners' ready to join the milking herd. Using AI wasn't really an option as the animals were in groups and when one was observed to be 'on heat' and ready to be mated, it was a major effort to get her in and tethered for the attentions of the inseminator. The alternative was to borrow, hire or buy a bull to do the job naturally. He was usually a 'beef bull', that is, his progeny would be reared for fattening not milk production. They were more docile than 'dairy bulls' and more important, usually produced small calves thus avoiding the heifers having difficulty with their first calf.

I decided to buy a bull rather than rely on hiring one. For some reason I decided I wanted a Galloway rather than the more conventional Aberdeen Angus or Hereford. A friend put me in touch with a prospective seller who farmed in the wilds of the north-east and in reply to my telephonic question as to the bull's capabilities (such as proven fertility) his reply, in broad Northumbrian dialect, was, 'He's a great stock-getter, but he tends to wander a bit'. Fortunately hailing from that part of the country I had no difficulty understanding either the dialect or his meaning. We did the deal and 'Oswald of Kircumputtick' (yes really) duly arrived to take up his duties.

One of the more obvious advantages of employing natural service is that the bull soon identifies the animals which are in season, or as farmers usually refer to 'on heat'. Otherwise one has to rely on behaviour patterns such as mounting and general restlessness – not easy to observe, especially in the winter. Oswald certainly seemed to have the nose for the job. This awareness was not limited to his interest to the animals in my care but extended generously to any which were within a couple of miles. Although one must never trust a bull, he seemed of placid temperament until he detected the telltale signs on the wind. Then he was off, head down through fences, over gates, across streams and into whichever group of female animals took his fancy. Many were the times when irritated neighbours demanded he be removed with utmost urgency. Fortunately he had a ring in his nose and the skill was to get the hook of the bull-pole onto the ring so that he could be led or pulled back home. Too often this was not quick enough to have prevented his achieving his aim. He certainly lived up to the reputation that he had gained in Northumberland.

Our expanding family

Our three sons were born in Cheshire. Wilfrid Peter in June 1964 during the time we were living at Bay Tree Farm. A large bouncing baby, we experienced nothing like the emotional roller coaster of Susan's birth twelve months previously. Unfortunately Barbara began to lose blood and the doctor decided she had to return to hospital. They, however, refused to allow her to take the baby and I was left holding it in one hand and Susan in the other. My brother Derek was due to come down to spend the weekend with us. He pulled into the yard as the ambulance left. He didn't bother getting his golf clubs out of the car. We managed competently, much to the amazement of our kindly neighbours. We even had Yorkshire puddings for Sunday lunch. Needless to say Pete has never looked back.

David James arrived in March 1966. This time, unwisely as it happened, it was decided that he should be born at home – by then we were at Spy Hill. The birth went OK but complications quickly followed and emergency procedures had to be employed as Barbara had again lost a lot of blood. Only Robin John born almost exactly two years later in Davenham Hospital, Northwich, arrived without the usual panic which seemed to surround these events.

So by the end of March 1968 we were the proud, if rather exhausted, parents of four healthy, mostly happy children. Life was extremely busy with the job, the farm, the family and even rugby in the winter months. There was much to do and little money to spend on the house to make it habitable. But the location was brilliant. Plenty of space, no neighbours within eyesight or earshot and a lifetime ahead to build the home of our dreams. Fate had certainly looked kindly on us so far.

CHAPTER FIVE

The Seventies

THE EARLY PART OF THE DECADE saw Ted Heath come to power, widespread industrial action and mounting troubles in Northern Ireland. Despite General de Gaulle's antagonism it also presaged Britain's entry into the Common Market. By the end of 1973 rocketing inflation and increasing unemployment led to the three-day week. It was a bad start. The October 1974 election was won by Labour with Harold Wilson forming a government, but with a tiny majority of three seats. Heath was ousted by Margaret Thatcher who quickly began to revitalize the Tory party. Callaghan (what crisis?) followed Wilson as Prime Minister and promptly lost the 1979 election when he unexpectedly resigned. Mrs T, Britain's first woman Prime Minister, set about the uphill task to economic recovery. Her first priority was to curb the powers of the trades unions, and the battle lines were soon to be drawn up with Arthur Scargill and the miners.

Move to management

I had been working as an MMB Consulting Officer for the best part of seven years and my main aim was to build up a reputation as a competent professional. Nevertheless I was fairly ambitious and the MMB was a large organisation with plenty of potential to progress. The main problem was that you had to be prepared to move about the country or work at Head Office in Thames Ditton if you wanted to get on. Our farm, therefore, now became a constraint. Whilst it would never be a significant income earner, it was obvious that the asset value of the land would continue to increase over the longer term. Besides, we were making it into a wonderful home with plenty of space for four children within which to grow up. Barbara was also now getting back into her medical career. Having done a spell in family planning and baby clinics for the local health authority, she began to do some locum work for a general practice in Chester. In the early eighties she was invited to join the Boughton Medical Group as a part-time doctor. She was to stay with this practice for over twenty years. So for these reasons any ambitions I might have had to be realised without a move from Spy Hill Farm.

Fortunately the demand for the LCP Service was increasing as dairy farmers were encouraged to expand their milk production through Government incentives and grants. They needed detailed business plans and their bank managers wanted reassurance that proper financial records would be kept and monitored. More trainee consultants were therefore being recruited and it became obvious that the Service needed to be properly organised and administered.

Initially the management structure had been fairly straightforward, being controlled centrally from Head Office. John Hodges, a senior man in the Production Division, had recruited me originally. He was tall, dark, rather suave, highly intelligent and with a good scientific background. He seemed to be the favourite to inherit the position of Director of the Division when the incumbent Dr Joe Edwards retired in a couple of years. Hodges was even dispatched to the Harvard Business School to prepare him for such an office. For some reason he didn't get the job and shortly after left the organisation to take a professorship at the University of British Columbia. From thence he worked for the Food and Agriculture Organisation in Rome. Norman Coward took over as Head of LCP in the mid-sixties, again operating from Thames Ditton. Norman had joined the MMB at the same time as me and was a Wye College postgraduate. An affable, bright individual with a sharp mind he was soon snaffled by the Midland Bank and eventually ran their specialist agricultural unit.

By the start of the seventies there had been various attempts to decentralise the management system of the Department and to amalgamate different groups of field staff under the administrative control of the MMB Regional Managers. But they had little understanding of the nature of the job and besides were only interested in making sure everyone filled out weekly reports and provided details of where they went and when. Their involvement added no value and in addition alienated most of the consultants in the process.

Change, however, arrived with a new director, Jim Morton, a large, jovial man appointed to head up the Breeding and Production Division. More a salesman than a scientist, Jim had been in a senior management role with one of the national feed suppliers. He understood farmers and he was determined to introduce a more commercial approach to the AI, Milk Recording and LCP Services within his empire. Furthermore he had good man-management skills and was able to fight his corner in the administrative corridors of power at Head Office. He was also determined that the Service would be managed by people who had a successful track record in doing the job themselves.

I was promoted into a management position and became one of five Regional Consultants. The head of the Department, by then Roger Hurrell, suddenly moved sideways to a public relations role in Head Office and before I had time to catch breath I was applying, with tongue in cheek, for his job, on the understanding I didn't have to move to London. With some amazement I got it and took over as head of the newly and rather clumsily named Farm Records and Consultancy Department (FMR&C) in April 1972. Although I didn't have to move I had to work from an MMB Regional Office. Fortunately the nearest one was only twenty-five miles away at Wrexham in North Wales. A farm records section had already been established there to deal with clients' data, although the office still operated as the administrative centre for milk producers in that part of the country. The Regional Manager, J.P. Jones, was a small, bald-headed Welshman who quite liked the idea of having the head of FMR&C based in his fiefdom. However, he was determined that he should still be seen as the senior MMB man in the area. Apparently there was no suitable office accommodation in the main building, which had been previously a rather large Victorian house in Grove Road, so I was relegated to the garden and into what amounted to a small caravan just big enough to house me and my secretary, Meg Morris. There was certainly no room to swing the proverbial cat. We even had plastic chairs. It wasn't too bad for me as I spent most of my time in different parts of the country, but for Meg it could be freezing in the winter and boiling in the summer. She was a wonderful person, patient, long suffering and efficient. Not only did she cope with all my work, but also much of that of the Regional Consultant, John Butler, who had taken over the West Midlands Region from me.

Jones was a rather introverted man with not a vestige of humour. We tolerated each other from a distance and he delighted in sending me long memos about some staff misdemeanour rather than walking a few yards from the office to the garden to have a word in my ear. He didn't last that long, being a casualty of rationalisation as the number of milk producers continued to fall and regional administration work for North Wales was amalgamated with the larger office in Newcastle under Lyme. When he left I took over his office in the main building, but as a parting shot he removed every piece of furniture including the carpets. I have no doubt that his views of me were not too complimentary either.

The Wrexham office was eventually to become the national administrative centre for the consultants and in addition the main postal dairy herd costing service (Dairy Management Scheme) for the whole country was run from there. The office was efficient and friendly so that relations with both field staff and clients soon began to improve. I continued, however, to fight battles with some of the MMB Regional Managers who had my staff operating in their areas. To be fair they were often the first port of call if there was a problem or complaint, but it was an uneasy relationship.

So, suddenly without any training, I stopped doing something I knew about and became a senior manager ultimately responsible for some 180 farm recording and consultancy staff providing on farm services to about 3,500 farmer clients throughout England and Wales. It was a daunting prospect.

Changing the culture

One of my first tasks was to reverse the disappointing trends of recent years in financial results. Unless the Department at least broke even, there was a real threat that the Board would decide to wind it up. Rampant inflation in the early seventies meant that costs were escalating and fees had not kept pace. All increases in fees for services had to be formally approved by the main Board and as farmers' representatives they were less than enthusiastic in agreeing to add to their cost burden. So I spent some stressful times in front of the main Board arguing that farmers had to pay more. It was a valuable introduction to MMB politics, a subject with which I was to become especially familiar in the years ahead.

Irrespective of the type of business, the basics are usually much the same. We had to maximise income through achieving high membership per unit of field staff at competitive fee levels, and costs had to be squeezed out where possible, but not to the extent that staff morale was sacrificed. And there was competition. The Government-funded Agricultural Advisory Service employed substantial numbers of field advisers who were available to farmers for free. Most of the large feeding stuff and fertiliser firms provided 'free' advice on the back of their profits so our staff had to be that much better if they were to retain their clients and attract new business.

Styles of management differ quite markedly. There are those who direct from the top with little debate. Fear is usually the motivation which drives their staff. It can be an extremely effective method, but not particularly pleasant for the people who work within it. Teamwork and consensus is the other extreme. At worst, this manifests itself as indecision or compromise and can quickly lead to meltdown. Somewhere in the middle is probably the ideal, although I am likely to be a bit too near the consensus end. The main priority for me was to identify and promote individuals with different skills but who shared the philosophy of a team approach. I had a head start with some of the recently appointed senior staff and after some comings and goings I put together a team of four Regional Consultants, each with a deputy to manage the business in their respective areas. We set about changing the culture of the organisation.

Our main objective was to provide clients with the best possible consultant. Mature people of real calibre were rarely on the market so we had to grow our own. We set about visiting the final year students in all the major agricultural colleges and university departments. All employers want the pick of the bunch so we had to sell the job, rather than expect them simply to queue up. Our interviewing process became an overnight two-day event. It involved group work, practical presentations and individual face-to-face sessions. We needed not only intellectual ability, but just as important, youngsters with enthusiasm, judgement and good communication skills.

The new recruits went through an intensive training period so by the time they were able to do farm visits themselves they were well prepared to provide real value for money. This was the only way to beat the competition and the free service from other sources.

There were, of course, downsides to such a policy. After a couple of years with us, these bright young things were highly sought after by companies who could pay much better salaries than we could afford – notably the banks who were rapidly expanding their agricultural teams. The temptation to 'go it alone' and take clients with them was always a threat, but in practice this only happened occasionally. Also the client having got used to a good consultant did not like change, sometimes resulting in his resignation. But overall I am convinced this was the right strategy and helped us to build a reputation for quality, commitment and practical application.

One of the group activities involving the consultants was their annual conference, usually held in the first week in September at one of the agricultural colleges: Harper Adams in Shropshire and Myerscough in Lancashire were popular venues. In the early days these had been pretty uninspiring affairs discussing the finer points of bull selection or the design of farm recording forms, although the opportunity to meet colleagues from other parts of the country was always a popular spin off. We decided that future conferences would be much more participative and outward-looking than had been the tradition in the past. We invited top speakers, case studies were worked on by groups until late in the evening with formal presentations to the whole conference and we organised outside visits on the Wednesday afternoon to venues of local interest, often unconnected with farming. We even played a local rugby team one evening. The atmosphere had the ring of our student days and the final dinner usually ended up with much beer consumption and raucous singing.

The consultancy job was a pressure one. You might be dealing on the same day with someone who was going bust or someone else who wanted to buy another farm – the responsibility was often onerous. Although each individual was a member of a regional team and met both his colleagues and his manager each month to discuss issues of the day, it was still essentially a lonely occupation. Working from home meant you might not see other consultants for weeks on end. Ways and means of alleviating that pressure through events like the annual conference had an important part to play in maintaining morale.

The responsibilities in my new role were not confined to the Consultancy Service. I also needed to be closely involved with the farm records side and the new service developments which were coming on stream to help farmers monitor their dairy herds and businesses more effectively. We had a Development Team based at Head Office, so I began to spend considerable time travelling down to London. Fortunately there was a good train service from Crewe which meant I didn't have to waste hours in the car with the inherent dangers of motorway travel, and I could also work on the train.

My key management team were the four Regional Consultants – Ron Higgin in the north, Alan Croft in the south-west, Steve Churms in East Anglia and the south-east and John Butler in the West Midlands and Wales. All had strong personalities and had been experienced consultants who had earned the respect of both their clients and colleagues. We were all in our early to late thirties, hugely enthusiastic and had a similar outlook on life. We worked hard and we played hard.

Ron, a former burly, front row forward, was from the heartlands of Lancashire. He also had a small farm north of Preston, where he kept suckler cows as a sideline. He joined the MMB before I did, as a Livestock Officer, in the early sixties. He was down to earth, practical and totally reliable. He had to face the tragedy of losing his wife at a time when his three children were very young. My father would have identified with him the problems he then faced.

Alan Croft was a very live wire. He was a graduate of Leeds University and had first joined the Board as an inseminator in the sixties. Ambitious and able, he worked his way up quickly to the Consultancy Service. He was always going flat out, leading from the front and challenging the status quo. He could be a bit of a handful. He too had purchased a small farm in Devon and even attempted to bottle and sell water from a spring on his land.

Steve Churms often reminds me of the story of his interview to join the Consultancy Service. Apparently I was on the panel and he only got in as a result of a first choice deciding not to accept the offer. Steve was a tall, thin man with a passion for West Bromwich Albion and Tory politics. He could also be a bit intense. His region was, in many respects, different from the others. Dairy cows were few and far between in East Anglia and this meant long distances between clients and the need for his team to know more about arable farming than those in the middle and western parts of the country.

I first met John Butler at Chester Rugby Club and we often played together and participated in annual tours to foreign parts. One such to Amsterdam could be the subject of another book. A Reading graduate, John had gone dairy farming with a friend in Cheshire as a tenant on the Bolesworth Estate. It soon began to dawn on him that the tie of dairy cows and the prospective returns were not going to fulfil his expectations and he was looking for other opportunities. My informal conversations with him about the consultant job came at just the right time and he joined the MMB. With his first-hand experience and ability to communicate at all levels he quickly made a name for himself.

There were other key players at this time who took responsibility for development, sales and marketing. It was a vibrant and exciting period and the business began to expand. By the end of the seventies we had over 4,000 clients on the books and the 1977 Conference photograph records more than seventy consultants and trainees. We were even beginning to make money.

The public image

As I became better known, invitations to speak at meetings and write articles increased significantly. These could be local farmer groups, commercially sponsored events or large conferences with a few hundred in the audience. We had established an Information Unit in Reading which received coded copies of full farm data from recorded farms and when analysed by financial criteria provided an extremely good source of comparison as a base for presentations and discussions. The range in profit between the best and worst performers was truly breathtaking. It provided excellent material for discussion. Although I didn't particularly enjoy public speaking and probably over prepared, it was an essential part of the job. I needed to be seen out and about.

I even made a number of live TV appearances. Dan Cherrington, a former contemporary at Newcastle had taken over the presentation role for the Sunday farming programme from his father, John Cherrington, a revered journalist and farmer from Hampshire. I knew Dan so well that I had entrusted him as my best man to steer me to safety at our wedding some fifteen years earlier. He was a total disaster and I ended up after the reception with a bride in one hand and a 56 lb ball and chain in the other! Of course, he was one of the infamous

'Newts' and to this day still blames me for leading him astray as a fresh-faced student. It was certainly interesting, if not nerve-racking, to be discussing various aspects of dairy farming with such a character as we were beamed into farmers' sitting rooms throughout the country. I never quite knew whether he would lob an impossible question at me out of sheer devilment.

I also remember another Thames TV programme on which I appeared at short notice. It followed on from an MMB AGM at which there was the usual hullabaloo about low milk prices. A journalist had made a film on a dairy farm somewhere in East Anglia about a farmer who was bucking the trend and thought milk prices were fine. A phone call had gone from the producer to the MMB to ask whether the Chairman would care to go on the show and comment on the film and the assertions this man was making. Needless to say the Chairman had better things to do and the ball kept rolling down the hill until it stopped with me. I duly went into the studio in London for the live interview. It was what is known in the trade as a 'down the line' interview, that is to say, the interviewer would be in Norwich and I would be in the studio in London. He called me just before the interview and we talked in general terms about the film which was to be shown and the sort of questions he would ask. I was fairly nervous about all this. I could hold my own about dairy farming matters but was woefully ignorant about the machinations of the politics surrounding milk pricing. What I didn't know before going into the studio was that there had been a bomb explosion at the Tower of London an hour or so earlier and understandably there reigned some chaos as news flashes kept interrupting programmes.

The green light came on and the programme began, starting with the film on the farm. Unfortunately there was a fault and I could hear no sound at all. When the film finished my interviewer appeared on the TV monitor and said 'we are now going to the studio in London to discuss the issues raised in the film with Dr John Craven, Head of the MMB Farm Management Service. I could hear him perfectly. The next picture on the screen I shall never forget to my dying day. It was me staring open-mouthed into the camera. I stumbled and stammered my way through an explanation of not having heard the commentary and answered as best I could the questions posed and then staggered thankfully to the pub next door to recover my composure. I never met anyone who owned up to seeing this programme but I have ever since had a deep respect for anyone who works in live television – anything can happen.

Research and development

Apart from farmer meetings there was also the wider dimension of agricultural research and extension. The MMB had an impressive record in applying scientific advance into farm practice. It had pioneered AI and bull-breeding programmes. It incorporated nutritional evaluations within rationing programmes for cows and was engaged in identifying improved methods of disease control through its Veterinary Services Department.

Agricultural research in the UK was and still is a massive undertaking. Modernisation of farm practice and the increases in productivity in my lifetime have been huge. Government-sponsored research institutes for dairying, animal diseases, grassland, mechanisation and many others churned out trial work and recommendations for farm improvement on a regular basis. In addition, the Government-sponsored Agricultural and Development Advisory Service (ADAS) and agricultural faculties at the universities contributed to the torrent of scientific papers, many of which were completely unintelligible to the average farmer. That is not to say that all research workers were far removed from farm practice – there were some outstanding individuals, but many others were a few steps removed from reality. It was for those working at the interface between research and farming (often described as extension workers) who had the job of translating the relevant science into practice. MMB Consulting Officers were ideally placed to contribute to this process.

When I joined the MMB in 1963 the Chairman was Sir Richard Trehane, a tall, impressive man with a rather refined accent. Although he often appeared aloof, he was undoubtedly very able and enjoyed an enviable international reputation. He had taken the chair in 1958 and retained it for nineteen years until he retired in 1977. He was a Reading graduate and had worked at Cambridge as a research assistant to Sir John Hammond, one of the leading animal scientists of the day. Their research in 1936 had been into the pioneering technique of artificial insemination in cattle. After returning to the family farm in Dorset he soon became involved in dairy farm politics which culminated in his election to the MMB. Trehane played a crucial part in influencing the development of the Breeding and Production Services Division with his intimate knowledge of both the science and practice. It was not plain sailing. There existed strong opposition to some of the Board's influence, especially in the cattle-breeding field. Pedigree breeders were often outraged by decisions about bull selection for the national stud as they saw it as unfair competition to their own self-interest. Trehane, however, was determined to present the balanced view so ensuring that the majority of commercial milk producers had access to the best possible resources. Furthermore he supported his staff publicly when he was convinced of the arguments they presented.

In the animal production sector of the industry, the means by which researchers published their work was through submission of papers to various scientific journals. One of the main ones was that produced by the British Society of Animal Production (later 'Science' being substituted for 'Production' in its title, or BSAS for short). The Society held its annual meeting over several days for presentations based on many of the leading topics of the day, and at which a wide cross-section of academics, research workers, advisers and commercial representatives met to discuss the merits of new advances in technology. Concurrent sessions for those interested in pigs, sheep or beef cattle provided specific interest to people working with different species and the conference came together for plenary lectures by leading figures in the industry. Trehane was keen to see MMB staff at the forefront of this activity when the subject involved dairy cow fertility, nutrition, disease control and herd management. Staff from the Breeding and Production Division were frequently on the platform or in the audience and over the years I presented a number of papers concerning management related topics.

I was later encouraged to join the committee of the Society and became the Liaison Officer to the European Association of Animal Production (EAAP) which had been established in 1949 and of which Trehane had been President for six years in the sixties. It seems he had me in mind to carry the MMB flag into the European arena.

European Association of Animal Production

EAAP was a much more complicated organisation than BSAS. It stretched far wider than the EU and included representatives from equivalent bodies in countries from both Eastern and Western Europe. Also some countries around the Mediterranean were members. The farming activities were, therefore, very broad due both to varying climate as well as the general economic conditions which prevailed across the continent of Europe. The problem of language was a constant headache for the organisers. There were four 'official' languages – English, French, German and Russian. Simultaneous translation was expensive and could only be afforded for the full plenary sessions, whereas many of the individual presentations in the specialist Study Commission sessions were in the authors' own language. The Russian authors had some difficulty getting their points across.

Each year the Association held a meeting in a different country. It lasted about a week and usually attracted about 500 to 600 delegates, many of who brought their wives (I took Barbara on several occasions) and an associated social programme was organised for them. In the evenings there were joint dinners, river cruises or visits to the opera. Whilst there was a serious scientific purpose to the working sessions, some would argue that the main value was in the forum of the meeting, allowing colleagues working in similar fields and under different conditions to meet and discuss their work. Others may take the view it was a bit of a jamboree and there were too many hangers on. The truth probably lies somewhere in the middle.

There were six Study Commissions; three were specific to different species: cattle (dairy and beef), pigs, and sheep (and goats). The other three were concerned with the scientific disciplines of animal production: genetics, nutrition, and animal health (and management). Each Commission had a President and a Secretary and it was their job to choose the subject matter and invite the main speakers to present appropriate papers. Usually there were some combined sessions where a species commission worked with a discipline commission to produce a combined programme.

I was invited, no doubt engineered by Trehane, to give a paper to the Cattle Commission meeting in Vienna in September 1973. The subject, as I recall, was something like 'Dairy Farming in a Maritime Climate'. It was a review paper outlining the current scene in Britain as far as dairy farming was concerned. Trehane was in the audience and was kind enough to say that 'there was plenty of meat in the paper' – whatever that meant. Within a couple of years I was appointed the Secretary for the Cattle Commission and worked with the President, a very large Dutchman called Professor Romart Politiek. He was a key research man at the University of Wageningen in The Netherlands and also had his feet firmly on the ground. His English was variable at best so I had to write his presentations and reports to the Council. He was a great enthusiast and good fun to work with. My job was to help plan the programme for the subsequent meeting, agree the speakers, contact them, often in far off places and arrange for their papers to be assembled and distributed. It was quite hard work on top of my day job, but an important one, especially keeping our own British Society on the front page as it were.

I served as Cattle Commission Secretary for six years and attended something like twenty consecutive annual meetings in different countries during my involvement. Most of the venues were in Western Europe such as Copenhagen, The Hague, Madrid, Lisbon, Toulouse, Brussels, Zurich, Munich, Dublin and others. In 1979 the EAAP annual meeting came to Harrogate and the British Society was deeply involved in the organisation. To have the opportunity to learn something of the farming conditions in Eastern Europe as well as to visit cities like Leningrad (before the lifting of the Iron Curtain), Warsaw, Budapest, East Berlin and Zagreb was a real bonus.

After my spell as a Commission Secretary I was invited to do some research on ways to improve the efficiency of administration of the annual meeting, which culminated in 'The Craven Report'. This was pretty tedious stuff about submission of papers, layout of visual aids and procedures for presentations so that some sort of guideline could be distributed to all participants in an attempt to make the whole thing more professional. The British Society had a high reputation in this field and I used their templates as a base for my proposals. I was then invited to join the EAAP Council during the eighties and it became clear that Trehane (who had by now retired from the front rank but still retained much influence) together with Secretary General, Jan Boyazoglu, had me in mind to take over the role of President of the Association and indeed I was appointed as Vice President. But my job at home was changing and I simply could not commit myself to take on this extra responsibility. In many ways I rather regret that but I'm sure it was the right decision. The British Society, however, was kind enough to elect me as their President in 1982.

There was even a World Association of Animal Production (WAAP); again one of the prime movers was Sir Richard Trehane. It held its inaugural meeting in Rome in 1963 and was organised on similar lines to its European counterpart but on a four-yearly basis. Trehane was successful in persuading the equivalent scientific societies in the USA, Australia, Canada, Japan, New Zealand and South Africa to become founder members. This, of course, massively opened up the agenda to include the problems faced in developing countries and the means to feed rapidly increasing populations. I was fortunate enough to attend meetings in Australia, Japan and Finland during my involvement with EAAP. Some of the speakers and presentations were outstanding and brought home to me the real challenges faced by those with limited research resources and yet problems of a much greater magnitude than ever we had in Europe. It was a humbling experience to have to try and understand, let alone reconcile, the reasons behind the massive food surpluses in the EU on the one hand, with appalling examples of mass starvation in some of the developing countries on the other.

Overseas consultancy

One of the most impressive speakers I ever heard was an Indian by the name of Dr Varghese Kurien. His paper at a WAAP meeting described a rural development programme in his home country entitled 'Operation Flood'. It had been started in 1970 by the Indian National Dairy Board, of which Kurien was Chairman. Financed through the sale of skimmed milk powder and butter oil donated by the EU, the aim was to create a nationwide milk grid.

Milk production in India is very different from that in Western countries. The principal animal is the buffalo not the cow. Most of the milk producers have one or maybe two animals and are to be found in the rural villages. There are literally millions of them. The buffalo is a multi-purpose animal, providing in addition to milk, draft for pulling carts and ploughs as well as a producer of fuel! Buffalo pats are collected, usually by the women, dried in the hot sun and used for fuel when temperatures fall. The animal produces a calf every couple of years and as with the cow, her milk production will peak after about a month and then tail off to a dribble over the following year. The village farmer may have very little land and these ruminant animals browse the surrounding countryside consuming anything which takes their fancy. Buffalo milk contains about twice as much butter fat compared to cow's milk, and as a source of food it is of great value to the local village community. Although the first call for the milk would be for the farmer's family, any surplus was sold to the middleman for onward sale to others in the village or to those living in neighbouring towns. These middlemen were invariably unscrupulous and mixed water with the milk to increase the potential of their sales. The farmer received only a small fraction of the value of his product.

The concept of Operation Flood was to create a milk producers' co-operative in local villages. It would receive surplus milk every day and a simple fat test would be conducted at the point of sale. The producer would be paid in cash on the spot at the going rate. The co-operative would then sell the milk on to the consumer. Any profits the co-operative made from the operation were used to procure inputs such as feed, fertiliser and medicines so that the farmers were able to increase the productivity of their animals. As the scheme began to gather momentum the co-operatives were also able to finance the building of village schools, provide wells with clean drinking water and generally contribute to a raising of local living standards.

Kurien and Trehane were great friends. In some ways the MMB concept of co-operation for the benefit of milk producers set the basics for Operation Flood. Obviously circumstances were very different, but the aim of ensuring that the primary producer got a fair price for his product was the same. The two men clearly had much in common and as the project progressed and expanded, they must have discussed how MMB staff might be able to help.

At the end of 1978, three of us were invited to go to India and report back to the Chairman. The most senior man was John Frappell, the head of the AI Service. John was a vet and knew as much about the reproductive capabilities of farm animals as anyone in Britain. Furthermore he led the operation of an AI service and breeding programme which had, at that time, an impressive worldwide reputation. He was also a delightful companion with a great sense of humour. The third member of our group was a geneticist called Sandy

McLintock. He was an expert in breeding and selection programmes.

India was the first developing country I had been to and although I was expecting to see poverty, the culture shock was severe. I also anticipated crowds of people, but the teeming hordes on foot, on bicycles, in rickshaws, on scooters or in ancient cars was mind-boggling. The sacred cows added to the general chaos and there seemed to be no driving rules at all. The Highway Code was not much use in India.

The Indians who were to show us how the project was working were young, intelligent and committed to their work. We were soon in the villages and, as is the custom, welcomed with garlands of flowers. The whole population would accompany us around the village and proudly point out the new things which they had acquired as a result of the milk co-operative success.

It was not just the agricultural benefits which were flowing through. Male and female from different castes now stood in line when taking their milk for sale – something which would have been unheard of prior to the project. Where village farmers would bring their sick animals in to see the local witchdoctor, the village now had a visiting vet who, with the use of simple antibiotics, soon convinced the farmers that they had been previously hoodwinked. Even a rudimentary AI service had been started in some of the villages and the education spin off for the local inhabitants in terms of a better understanding of reproduction was a real, if indirect, benefit.

We were in India for three weeks, only sufficient time to get a superficial impression of what was happening. We learnt a great deal and there were things we could point to from our own collective experience which might be of benefit to them. But this was not a case of recommending Western dairy farming technique. Others had tried this approach by advocating the importation of specialist dairy cows so that milk yields could be increased substantially. But under tropical conditions, heat and disease problems rendered such a policy, at the time, unworkable.

As might be expected there was some criticism of Operation Flood within India. Apart from the foreign breed importation attempts, some of their research scientists and politicians argued that too many resources were being directed to milk production rather than to other crops such as grain which would have given a more productive return to help feed the millions of poor people. I'm not qualified to comment on the macro-economics but to me India had been an extraordinary experience. It was going to take time, but I had no doubt that with the enthusiasm and skills of the Indian Dairy Board staff, the project would be a great success. It is no surprise that India, like China, is now beginning to make such a major impact in the world economy.

A couple of anecdotes are worth mentioning on this trip. The first was the need for us to board a local bus as our private minibus had broken down and we were miles from our next destination. I can truly say that this journey of about two hours was the most hair-raising of my life. I knew it was going to be different when we had to help push the ancient machine in order to get it started. John and I were given the honour of the two front seats and then everyone else piled on. The pictures you see of people hanging on to the roof for grim death and hens flying about are not far from the truth. We set off as dusk began to fall. The road was hard surface in the middle and was the width of the bus. At either side there were earth tracks each as wide as the road. These were busy with camels, bullock carts and workers returning from the fields. Our driver seemed to be in a tearing hurry and set off at an alarming pace down the middle of the road. Occasionally another bus or more often a heavily laden lorry would be sighted in the distance. There then followed a game of chicken. The bus driver would career up the middle of the road with the intention of intimidating the oncoming driver to seek refuge on the hard shoulder. It didn't always work and at the last possible moment we skidded off the road and missed death by a matter of inches. After half an hour of this, John and I sat on the floor with our backs to the window – we couldn't take any more.

In the middle of the trip, we attended a meeting at the Dairy Board's Head Office at Anand not far from Ahmedabad, a large sprawling city in Gujarat State. It was exceedingly hot and unfortunately we were in what they chose to call a dry state: that is to say no alcohol. Making our unhappiness known to various people, we had a visit from one of the staff bearing a large box containing an assortment of spirits, beer and wine. He produced a form for us to sign, which upon inspection turned out to be a statement that the consignment was for registered alcoholics. We signed the form without hesitation. No doubt I am now listed as an undesirable alien on one of their computers.

I returned to India as part of another team a couple of years later, this time financed by the Commonwealth Office and including a cross-section of expertise with a wider remit. It is a fascinating country with so many reminders of colonial rule mixed with a rich history of its Mogul empire. We even managed to squeeze in a visit to the Taj Mahal, truly a wonder of the world.

I undertook two other overseas consultancy jobs whilst working for the MMB. The first was in 1969 to South Africa at the time the Americans landed on the moon. The Natal Milk Producers Union had started a dairy herd costing service based on LCP and one of our Farm Recorders had been seconded

1. My father, Wilf, aged about 10 (seated left) with his three brothers, Norman, Reg and Harold, 1913

2. Hilda and Wilf, 1928

3. 'The Chesters', Stanley. Wilf and Hilda's first home, 1930

4. Wilf and Hilda's wedding. Wilf's parents, Thomas Cass and Sarah Jane are standing on the left Hilda's parents, Jesse and Joe Walton are standing on the right

5. In siren suit with Sheena, 1939

6. Start of a lifelong addiction, 1939

7. With Mother, Father and Derek, 1940

8. Holmslyn, Whitley Bay, 1944

9. Holmslyn, 2011

10. *Mowden Hall School, Stocksfield, Northumberland*

11. *Mowden Hall Rugby XV - aged 11 and standing fourth from the left*

12. Durham School and Cathedral, 2011

13. Durham Regatta, record haul, 1954 - Standing on the left

14. Durham University VIII at Chester Head of the River, 1957
I am at no.5, counting from the front of the boat

15. Durham University Rugby XV 1960/61 - hirsute captain

16. *University Agricultural Faculty at Newcastle, 1956*

17. Professor Mac Cooper (right of picture) Dean of Agriculture, 1954-71

18. *HTY 674 on tour in Spain, 1958*

19. Pele Tower at Cockle Park, 1962

20. Dr Barbara Cooper, 1961

21. John Craven, an unlikely suitor, 1961

to them for twelve months in order to help them get it going. They needed assistance in calculating the annual summaries of their 100 or so members and asked us for advice. I went out on a six-week assignment and worked with our Recorder and their consultant, Brian Sugden. It was my first trip to a new continent and I relished the chance to see how dairy farmers in a different part of the world managed their enterprise. My base was in Pietermaritzburg about 100 miles from Durban but much higher on the veldt. It was extensive farming as land was not a limiting constraint. Labour was also plentiful and some very large herds were still hand-milked by coloured workers. In fact I was offered a job over there and it so happened that we won a lottery at the rugby club which enabled Barbara to fly out to join me for a couple of weeks. Apartheid was still very much in operation and we decided the political risk was simply too high to bring out our family and leave our relations behind.

I also joined a multi-national consultancy project in 1977 in Belize, formerly British Honduras in Central America. The country didn't have a lot going for it. The project, financed by the Commonwealth Office, was to consider the feasibility of Belize as a possible site for setting up a massive dairy cow enterprise to supply milk and milk products not only within that country but to other Commonwealth countries in the Caribbean. At the time of our visit there was a serious row going on with neighbouring Guatemala, which maintained that they should have a land access corridor through Belize to the Caribbean Gulf. We were billeted in the bush at a place called San Ignacio and the British CO of the local garrison arrived on our first night to warn us that if the balloon went up we should endeavour to be on the right side of the River Belmopan as their first job would be to blow up the bridge. As, no doubt, a result of the Harrier jets screaming overhead, the balloon kept firmly on the ground and we proceeded with our project.

It was not a success. Belize is a poor country about the size of Wales and mostly undeveloped jungle. The main output in colonial days had been mahogany and the British had not left much of a legacy. We did visit a Mennonite community who had developed about 5,000 acres of productive land. A religious group spurning modern technology, their forebears were people originating from Europe who had sought their fortune in North America. Over the years some of their sects had moved as far south as Central America. Their farming, although traditional and un-mechanised showed what could be done given the will. But the difficulties were profound and the West Indian colleagues within our group were against the project from the start. Our report no doubt still languishes in a dusty file somewhere in the bowels of the Commonwealth Office.

I celebrated my fortieth birthday in Belize. Naturally we had a party and enjoyed much local food and drink. Earlier in the evening I managed to get through to home on their intermittent telephone system. I talked to everybody but they all seemed so far away. Barbara later told me she had some tearful small boys to put to bed that night.

Foot and mouth disease

In 1967 we had a devastating outbreak of foot and mouth disease in Cheshire, which indirectly altered the farm enterprise. It also put an end to any farm visits and I had to work from home for some months, mostly doing calculation work on annual summaries which were mailed to me from Head Office for clients throughout the country.

The outbreak began in Shropshire in the autumn and spread rapidly north-east on the prevailing wind. Over 1,000 farms were affected and the disease lasted officially for 224 days. Total casualties were 90,000 cattle, 16,000 sheep and 42,000 pigs. The cost measured in money came to many millions but some would say the human cost was higher. Infected dairy herds carefully and lovingly bred over generations, together with the rearing animals and even the young calves had to be destroyed and incinerated on the farm. Many families never got over the trauma.

At this time we were contract rearing dairy heifers at Spy Hill. Every day I expected to find symptoms of the disease when I inspected the cattle, but fortunately we escaped, as did the owners for whom I was rearing. Poor old Oswald, my Galloway bull, ended his days during the epidemic. I had loaned him to my friend David Dodd to serve some of his heifers before the outbreak and he lost his herd and with it, the 'grand stock getter' from Northumberland. I doubt my neighbours shed many tears.

Compensation for the loss of the cows and consequential loss for interrupted milk supply was paid out by the Government as rapidly as possible to those affected. Some decided not to re-stock with dairy cows but find another enterprise with which to continue their farming without the demands of a 365-day commitment. Those who planned to continue producing milk had to find suitable replacements and the cattle dealers had a field day. This was a time when milk producers throughout the country were leaving the industry, not only because of its labour demands but also because it was becoming necessary to invest in modern milking parlours and buildings if they were to remain competitive. The capital cost for many was too high. The value of dairy cattle increased as demand from the foot and mouth affected areas intensified. It provided a golden opportunity for those who had considered 'life without being tied to the cow's tail' to cash in on the market. The time also coincided with the increasing and more widespread incidence of the highly contagious disease, brucellosis. Affected animals aborted their calves before full term, ruining the cow's lactation and providing a ready source of re-infection to their herd mates from the discarded foetus and membranes. The disease was also transferrable to humans as undulant fever, an unpleasant and long-lasting condition but not causing abortion in females. Those working closely with cattle were susceptible and a number of herdsmen became victims.

There is no doubt that a significant number of cows imported into Cheshire following foot and mouth were brucellosis carriers and there followed an epidemic of brucellosis hard on the heels of the previous foot and mouth devastation. Foxes and even birds are known to spread this affliction and sure enough some of our contract-reared animals became infected. This finally put paid to that enterprise and we had to find an alternative which could be managed with limited time and investment.

Beef farming

The answer seemed to point to beef production. Beef has an image of a subsidiary enterprise, especially on dairy farms. There are, of course, many specialist and profitable producers, but like an iceberg, there is much hidden beneath the surface. Britain eats a lot of beef. It has a long tradition based on the popular roast beef and Yorkshire pudding Sunday lunch. The French even call our rugby team *Les Rosbifs*, maybe because we refer to them as 'frogs'.

A high proportion of our beef calves come from our dairy farms. In a 100-cow herd of Friesian or Holstein (the black and white ones), the farmer would probably inseminate fifty of his best animals with a bull of the same breed which has a proven record for improving milk yields. About half of the resulting calves would be females, which he would rear as replacements for his dairy herd. The other half would be male and he would either sell these in the local market or keep them to fatten. Either way they would end up as beef-producing animals. The remainder of his dairy herd might be inseminated or served naturally by a specialist beef bull. This practice is known as cross-breeding and the resultant calf is much better suited to beef production than its pure-bred half brother. Both males and females from cross-breeding end up in the beef sector – so in this example three-quarters of all the calves born to the dairy herd go into the beef production cycle.

Another major source of beef calves are suckler cows often kept on the more marginal or hill land. This is a term given to a beef cow (usually a cross-bred dairy beef animal herself) which is put in calf to a beef bull. The calves are left with their mother for up to six months and thrive on a diet of milk and other feeds. They are then weaned and either kept or sold to other farmers for finishing. The suckler cow in the meantime is now pregnant and will produce another cross-bred beef calf, and so the cycle continues.

I decided against having a suckler cow herd. The critical issues were to get them in calf regularly and to ensure they calved successfully and with minimum hassle. If you lost the calf you lost the whole year's income. Both these priorities required careful attention and with my job I simply couldn't guarantee to be around. The alternative was, therefore, to buy calves and fatten them for beef. I chose to buy weaned calves from suckler cows at about six months old. Many would say that to make money out of beef you have to buy and sell at the right time and at the right price – the bit in the middle was less important. Ron Higgin, my Regional Consultant colleague in the north was an expert in these matters and often went with me to suckled calf sales in places like Skipton and Kendal. He gave me the nod when a pen of calves (usually six or eight were sold together) was a good buy and I bid while he went in search of the seller to extract 'luck money'. This was usually a pound or two per head in cash if the seller was pleased with the sale. Needless to say this long established tradition wasn't shared with the taxman.

When we eventually got them transported home, there would be an unholy row for a couple of days as the calves had not only been through the trauma of the market but also weaned from their mothers only a day or two earlier. It seemed a harsh process but within a week they had settled down in their new surroundings. They were purchased in the autumn sales and fed on a diet of rolled barley, brewers grains and straw during the subsequent winter. We already grew the barley so the production costs were low. They didn't put on much weight until they went out to grass in the spring and then the risk was that they would get too fat.

We had about eighty to one hundred animals each winter and I even had a local butcher slaughter some of them for sale to friends as quarters or half sides for the freezer. It was excellent beef, although we didn't make much money. That is one of the main problems in beef production – money is tied up in the animal until sale and then reinvested almost immediately to purchase its replacement. There never seems to be any ready cash available.

I was fortunate to have a couple of friends from the rugby club who liked nothing better than to come down to the farm on a Sunday and help out. This might be carting bales, mending fences or chasing cattle. Barbara would put on a splendid tea and their children could rush around and play with ours. Another friend and later an MMB Consultant colleague, Geoff Coupe, lived about half a mile from Spy Hill and ran a successful dog kennel business. Both he and another local neighbour were always willing to lend a hand and even the children were old enough to be dragooned into helping for a short spell – so I could keep labour costs to a minimum. Feeding the cattle in the winter took me about forty minutes before and after I returned from work, but Barbara spent more time than I would have liked standing in when I was away. Needless to say most weekends were busy catching up on fertiliser spreading, mucking out, drenching cattle or other of the myriad of tasks which needed to be done. At least I had given up rugby by the mid-seventies so I had a bit more of the weekend to devote to the farm.

Children's education and goodbye to my father

The children were now finishing primary school in the local village of Ashton Hayes and we had to decide on their future education. The secondary school in Tarporley had recently been launched as one of the new comprehensives, but we were disappointed by the standards and the sports facilities were practically non-existent. The picture is very different today, but timing is of the essence and we decided to send them to a boarding school, although they all attended Tarporley for a short while from the age of eleven. My old school in Durham was one possibility but it was 200 miles away and we wanted to participate in school events as much as possible. The Assistant Headmaster at Rydal School in Colwyn Bay, North Wales was an Old Dunelmian and having been shown round the school we were impressed by the facilities and decided to go for it, accepting that we had to give the same chance to each of our four children. We also knew the cost was going to be a substantial burden for many years, but we had two salaries and quite a bit of collateral in the farm. It was a big risk. It would have been a major setback if something happened and we couldn't meet the costs and had to withdraw some or all of them.

My father died in February 1977, shortly after his seventy-fourth birthday. He had lung cancer and had been ill for some time. I have no doubt that his pipe was a major contributory factor in developing the disease.

He was a much-loved man of great principle and integrity. Losing his wife at such a young age must have been a shattering experience, especially in view of the responsibility of bringing up two boisterous boys. He always maintained he would never remarry as no one could replace Hilda. In fact he did marry Lillian in 1967 some twenty-one years after she came to live with us following my mother's death. I was delighted that she became my stepmother.

My father had not had much opportunity to participate in sport as a young man, but he followed Derek's and my exploits on the river and on the rugby field with great interest. There was nothing he liked better than meeting our friends and providing much appreciated hospitality whenever the opportunity arose. He had a wonderful sense of humour and was apt to repeat family anecdotes to anyone who would care to listen, more often than not with tears of laughter running down his cheeks well before he got to the punch line. Derek and I had fictitious numbers for each of these stories and as he commenced we would declare loudly that this must be 24 or 36. He took no notice.

He and Lillian came to visit us in Cheshire quite often and delighted to see our children growing up on the farm and even beginning to play rugby.

He had a huge influence on my life and not many days go by, even now, when my thoughts or dreams do not feature him in some way or other.

CHAPTER SIX

The Eighties

THE POLITICAL SCENE in the eighties was dominated by Margaret Thatcher, Britain's first female Prime Minister. She faced a tough economic challenge as the trades unions were in militant mood, demanding wage increases to offset high inflation. She was determined to pursue policies which deregulated the financial sector, increase the flexibility of labour markets and close down state-owned companies. Understandably her honeymoon period was short-lived as the country sank further into recession. Had it not been for the Falklands War, she may well not have survived the election in 1983. She did, however, survive an assassination attempt the following year when the Brighton bomb exploded at the Tory Party Conference. Nicknamed the Iron Lady she took on and defeated the miners and stood up to the Soviet Union in the war of words. She won a third term in 1987 but the Poll Tax caused a great deal of civil unrest and her views on Europe were not shared by many in her Cabinet. The decade had its full quota of conflict, starting with the Iran/Iraq war and ending with the Tiananmen Square massacre in China. AIDS became a major health issue and DNA testing first featured in criminal identification. Perhaps the most significant change came in Russia as Gorbachev initiated democratic reform and the Iron Curtain began to fall.

MMB politics and milk quota

Trehane had been succeeded as Chairman of the MMB by Stephen Roberts in 1977. The two men were totally different in character, outlook and political appeal. Steve, as he was popularly known, had left school at fifteen to join the family haulage business and work on his home farm in Shropshire. He was a self-made man, loud, energetic and a powerful public speaker. He lost an eye when in the Home Guard during the war and had a glass replacement which wasn't very well aligned. You were never quite sure when talking to him which eye was the functional one. Trehane, on the other hand, was more representative of the landed gentry. He had a scientific training and an agile mind but it is possible that Roberts may, at that time, have had a wider appeal to the 40,000 dairy farmers left within the industry.

Perhaps Roberts' greatest claim to fame was his determination to expand the Board's milk processing creameries, and following the acquisition of new businesses (uncharitably described by some commentators as based on 'back of the envelope calculations'), he re-structured the organisation under the brand name of Dairy Crest, now a successful plc in its own right.

His lowest point came with the introduction of milk quotas. By the early eighties the milk product surplus in Europe was of the order of ten million tonnes per year and the cost to the taxpayer was unsustainable. It had to be either savage cuts in the guaranteed milk price or a draconian system of controlling total milk production throughout the EU. Although UK dairy farmers may have been able to withstand a price cut, there was no way that this would have been tolerated by the smaller and more militant milk producers in France and Germany. Roberts famously declared in March 1984 that quotas were completely unacceptable to British dairy farmers and they would only be introduced 'over my dead body'. They came like a bolt from the blue one month later in April.

By now Britain had effectively surrendered sovereignty of her agricultural industry to the bureaucrats within the EU. Furthermore, the UK government seemed disinclined to even argue the case for the British farmer, despite the National Farmers Union pointing out the iniquitous basis for deciding the national quota allocation based on previous production levels rather than the size of the domestic marketplace. The UK was actually a net importer of milk products and therefore hardly a subscriber to the surplus problem. The EU argument, of course, was that we were part of the Community and therefore must play a full part in solving the overall problem.

In any event the result was that British dairy farmers received an individual quota which was 9% lower than their previous year's production and if they exceeded it they had to pay a punitive super levy, way above the cost of production. This only became operative if the country as a whole exceeded the national quota it had been allocated, so the whole process soon became a major management issue with the MMB publishing weekly statistics of actual production compared to the quota profile. Farm advisers and consultants quickly devised milk monitoring schemes for individual clients so that they could gauge accurately whether they might exceed their quota by the year end in March.

There were hundreds of special cases, for example, where a farmer had just started production and had no previous basis for assessment. Special Quota Tribunals were set up to investigate each one and it took many months before these were finally agreed and the farmers concerned could return to something like normal life.

Perhaps the most significant development was the sudden emergence of a market in trading quota and its transference from one farm to another. There were special rules which applied, but effectively this meant that quota assumed a monetary value and all eligible producers suddenly found that they had a substantial additional asset (for which they had paid nothing) on their balance sheets. The whole process also provided an opportunity for quota brokers to set up as middlemen putting potential buyers and sellers in contact with each other. At the height of the demand in the early nineties the quota purchase price increased to a staggering 80p per litre, over four times the actual sale value of the milk. Even the annual leasing price peaked at 12p.

In 1984 quota values were much lower but for an average size herd awarded say a quota of 400,000 litres, the farmer could reasonably conclude that if the prevailing price of quota for sale was 30p per litre, then the equivalent asset value of his award was some £120,000, well over double the value of his herd. It was a hypothetical number as he realised nothing unless he sold his quota and with it, his eligibility to produce milk economically. Nevertheless, this sudden windfall rather dampened the outcry that had raged when quotas had been introduced a few months earlier. Some even maintained that quotas were a great idea. It was certainly a godsend for those who had already decided to quit the industry as they could now sell their cows and their quota. They could even get rid of their cows, retain their quota, then lease it out to someone else and receive a steady income for doing nothing. Unsurprisingly, milk producer numbers continued their inexorable decline. Whatever the national or individual reaction to quotas they worked and milk production throughout the EU began to fall back quickly to manageable proportions.

Quotas also enabled a young former consultant to make his mark within the MMB. His name was Steve Amies, a Reading graduate who had joined the LCP Service straight from university as a Trainee Consultant. He was a rather studious young man, small build and thin on top. One might have imagined him more at ease in an academic or research role rather than on a farm wearing wellingtons. It soon became clear, however, that Steve had an outstanding intellect and after a few years in the field was appointed to head up the LCP Information Unit at the Regional Office in Reading. This had been established in order to analyse collective financial data from our farmer clients and the results were of considerable importance to senior MMB staff when negotiating milk prices with the dairy trade, so much so that the MMB funded the cost of the Unit directly. Another benefit arose in the comparative studies which the Unit carried out between different EU countries with regard to milk production. Investigations in Holland, Denmark, Germany, France and Ireland were all undertaken by individual consultants seconded to the research project and overseen by Steve Amies. Their reports were then presented in person to full meetings of the main Board.

Steve duly moved on after a few years and became Personal Assistant to the Managing Director of the MMB. When quotas were suddenly introduced in 1984 a great deal of the work in explaining the system to farmers fell at Steve's door. He was the main spokesman to the press, wrote dozens of articles, spoke at farmers' meetings up and down the country and also did most of the original thinking as to how dairy farmers might make the best of the new situation. I still saw him frequently as he was based at Head Office but I didn't realise at the time that within a few years we would be working very closely together.

MMB politics were particularly active in the mid- to late eighties. Managing directors came and went in quick succession. Also at this time there had been speculation about an internal candidate by the name of Detta O'Cathain, then head of the Milk Marketing Division. She was Irish, capable and outspoken. For some reason, probably because she upset too many Board members, she didn't get the top job. She subsequently went off as MD to the Barbican Centre and reputedly introduced a reign of reform and terror before being ennobled during John Major's administration in the nineties. Maybe she would have been the answer.

Steve Roberts, the Chairman of the Board, retired in 1987 and had been replaced by Bob Steven, a dairy farmer from Kent. There had allegedly been a great deal of competition for the job within the Board (they elected their own chairman) on this occasion and no obvious candidate had emerged. Bob, a large affable man, seemed to me to be a compromise choice between various opposing factions. Individual ambitions within the Board seemed to be causing increasing unrest leading to muddled decision-making. Cracks were beginning to appear. Eventually Charles Runge became the new Chief Executive with a brief to prepare the organisation for deregulation. It was clear, under the current Thatcher policy of freeing up the market, that the days of the Milk Marketing Board's statutory monopoly power were numbered.

Charles was educated at Eton and Christ Church, Oxford. He worked for Tate and Lyle following university and was seconded to merchant bankers in London and New York before becoming Tate and Lyle's first Director of Corporate Affairs. He was bright, impetuous and became bored very quickly. He had a view on everything and usually a strong one. I remember him inviting me to join him as a guest at the 'Varsity match at Twickenham early in December one year. It was in the days of small corporate lunches of about twelve in a private room with access to an outside balcony in the old South Stand. Before long he was expounding the finer points of the game to some rather bemused former internationals in our party. He might have played the Eton Wall Game but had little understanding of the finer points of rugby, the game he had come to watch. But he was good fun, if exhausting, and I enjoyed his company. He was determined to introduce major changes to every aspect of the MMB. He also wanted to be involved with the MMB Dairy Crest business, but by now this had its own separate Board and Chief Executive. Runge was kept at a safe distance, much to his increasing frustration.

Change and challenge in the Breeding and Production Division

Meanwhile at the next level down there had been little change at the top since the early seventies when Dr Kevin O'Connor succeeded Jim Morton as Director of the Breeding and Production Division. Kevin had completed his PhD at Newcastle a few years ahead of me and had joined the MMB to work on the statistical intricacies of bull testing programmes. He had been a steady hand on the tiller for over ten years. Cool-headed and scientific in his approach, he was intent on expanding and improving the quality of the various farm services under his control. This was a perfectly reasonable ambition but times were changing and with Runge's arrival, both he and the Board began to question why we were importing so much foreign dairy bull semen and why our cattle-breeding programmes seemed to be under continual criticism by farmers and breeders.

O'Connor's empire, soon to be re-named the Farm Services Division was a considerable size in the mid-eighties. With a total fee income of over £30 million and staff numbers throughout the country just short of 1,500, the Division was a significant activity, especially as it also had dozens of farmer committees throughout the AI and milk recording regions. However, it was running at a loss. The principal problem was the cost of official milk recording. Because the bulls used in AI had traditionally been selected on the basis of individual cow records, it had long been accepted that the cost of milk recording should be subsidised by all producers to the tune of 30%. When AI profits had been booming, the profit from semen sales more than compensated for the net deficit in milk recording. By 1987, however, the position had deteriorated and the net result between the two activities produced a deficit of £1 million.

Nevertheless National Milk Records was a vitally important activity in that about half the herds in the country subscribed to it. It was expensive to provide as it involved official farm visits every month to carry out the work and in addition to take a sample of milk from every cow, which then had to be transported to the laboratory for analysis of fat and protein content. Detailed records were then returned to the farm within a few days. It was a massive logistical exercise.

Most farmers used the information to help manage their dairy herd. Decisions to cull or select individual cows from which to breed were impossible without reliable records. They were also used for feed rationing, monitoring herd fertility and disease incidence. The amalgamation of the information and the provision of daughter records for individual bulls on test were equally vital for those organisations like ours which were in the breeding business. Whilst the NMR subsidy had to be removed, it was entirely reasonable that those organisations which relied on this collective information should be prepared to pay for it.

There were three main departments within the Division. John Frappell (my former consultancy colleague in India and fellow registered alcoholic) had previously been head of AI but had retired and been replaced by Dick Shaw, another vet. I was head of Farm Management Services which, by then, included NMR, although in practice it was still run by Graham Nicholls who had been head of NMR for years. The Farm Management Records and Consultancy arm continued to thrive with a fee income of £3.5million and over 6,000 members or 18% of all dairy farmers on its books. The third department was Veterinary Services run by James Booth, also a vet. James was well known internationally as a specialist in mastitis control. He ran the laboratory in Worcester which undertook bulk milk testing for all the dairy herds in the country. This test identified the presence of brucellosis or the degree of severity of mastitis within the individual herd. In addition there was a wide range of other herd health-related services such as pregnancy testing, as well as the provision of an annual milking machine test carried out by technicians visiting farms. It was a small, highly specialised department, efficiently managed and invariably covering its costs through membership fees.

Kevin O'Connor was due to retire in July 1989 and the inevitable question arose as to whether I should go for his job. But this time it wasn't realistic to even apply for it without being prepared to move to London as all such senior appointments were based at Head Office in Thames Ditton.

The children, by now, were all in their twenties. Sue and David were both married and had their own homes. Peter was working in South Wales and

Robin was about to leave for New Zealand to seek his fortune. So the excuse that we couldn't leave the children was no longer valid – they were grown up and had already flown the nest. After much soul searching we decided I should apply and we came to terms with the fact that we would have to sell the farm and buy a property somewhere in the Thames Valley and not too far from Head Office. It also meant that Barbara would have to give up her part-time GP job in Chester. I didn't really expect to get it as I thought Runge would look for an external candidate. I undertook various in-depth interviews with external consultants, which further reduced my expectations but in November 1988, I was offered the job as Managing Director Designate to succeed Kevin O'Connor when he retired the following July. The question of location had not even arisen as I assumed Runge must know that we were prepared to move south. It was with some amazement therefore, when we did get round to the subject, that he said he wanted me to relocate the whole of the Divisional Head Office staff away from Thames Ditton. He even mentioned that we should find somewhere in the middle of dairy farming country and with good communication links. How about somewhere like Crewe in Cheshire? I nearly fell off my seat.

In the short term I was officially based at Head Office and in practice this meant commuting from home to London most Monday mornings. Occasionally I drove or took the shuttle from Manchester if I had an early meeting, but normally I went by train. I caught the 7.45 a.m. from Crewe to Euston, tube to Waterloo, suburban line to Surbiton and taxi from the station. If the trains were on time I could be sitting in my office by 10.00 a.m. I stayed in a local hotel during the week and usually managed to get home late afternoon on a Friday. It was far from ideal but both Barbara and I realised that the benefit of eventually being able to stay at Spy Hill was well worth it. Also it wasn't the same every week. I was often travelling round the country and Cheshire was central to many destinations. On one occasion, just before Christmas 1988, I was flying from Northumberland back to London at the same time as the Lockerbie air disaster occurred – we were probably no more than sixty miles apart. A sobering thought to remind me of the risks and perils of transportation in the modern world.

Premier Breeders

The debate concerning the MMB AI Service was by now beginning to gather momentum. The MMB breeding objective had been to steer a middle course between increased milk production and a beefier type of cow that would produce a male calf suitable for fattening: a typical British Friesian. This had suited most customers, whilst those wanting more extreme dairy-type cows purchased semen from Canada, for many years the only significant country from which exports had been allowed. Import restrictions were ostensibly for animal health reasons, though inevitably also carried a trade restriction element.

Whilst the MMB had been steering its middle course, countries like the US and Holland had been selecting flat out to improve the genetic potential for increased milk yield of the black and white Holstein cow. In the late eighties, with the animal health restrictions eased, US and Dutch genetics started to come into the UK. To protect our business, we imported this semen ourselves and sold it on to our AI members, making a useful profit in the process. An initial trickle soon became a flood, as farmers eagerly sought bulls to breed daughters that would produce more milk.

Whilst our producers were obtaining the genetics, it cost more than it should and the reputation of our breeding programmes was rapidly going down the pan. To be fair the earlier semen import restrictions had limited the progress of our own breeding programmes. This, coupled with fact that it takes five years to 'prove' a new bull, meant no quick solution was possible from within. However, the punters weren't in the mood for excuses, they wanted action and what was the Board going to do about it?

An independent, privately-owned cattle breeding company, based in Northumberland, called Premier Breeders had started a national Embryo Transfer Service in 1982. This was a technique which involved treating the cow with hormones so that she produced multiple eggs instead of just one, or occasionally two, when she ovulated. She was then inseminated with semen from a top bull and the fertilised embryos collected from the uterus and implanted in other recipient females for development and eventual birth of the calves. This had great significance in breeding terms as the best cows could produce many more calves within their lifetime than would have been the case with conventional AI. In addition this company had embarked on an ambitious venture aimed at establishing the nucleus of a herd of exceptional genetic merit from imported embryos from all over the US. The herd was given the prefix MOET, being the acronym for 'multiple ovulation embryo transfer'. The aim was to use 'sibling testing' (a comparison between sisters rather than daughters) as a quicker alternative to progeny (daughter) testing to identify outstanding bulls. In theory it was exciting stuff but it was also complex, untested and expensive. It was clear that the company were struggling financially and their bankers were probably getting very nervous.

Charles Runge had heard of this company and although unfamiliar with the finer points of reproductive physiology in general and cattle-breeding programmes in particular, was impressed with the new technology and convinced by their sales pitch that this was a solution to our problem. Kevin O'Connor, on the other hand, whilst accepting that embryo transfer was here to stay was more dubious about the concept of 'sibling testing', as the comparisons in statistical terms would be very unlikely to challenge the internationally accepted bull proving methods. But he was within sight of retirement and Runge's enthusiasm carried the day. He put the proposal to the main board and Premier Breeders was bought lock, stock and barrel for between £5 and £6 million pounds. At least this was action and the press had a field day. Their Chairman, John Moffit, a well-known cattle breeder, and their Chief Executive, David Storey, featured in the media pulling a cork from a bottle of Möet champagne.

The deal included two farms in Northumberland, one with a modern embryo transfer laboratory, another much bigger unit with a milking set up to house over 300 cows, all the cows, young stock and recipient animals, the associated staff (about thirty in all) and finally the negotiated goodwill. The latter represented the accumulated theoretical value of the whole project and comprised by far the greatest cost to the buyer. Not all commentators supported the purchase. Some of the less charitable suggested that the MMB should have waited until Premier Breeders had gone bust and then bought it at a fraction of the cost from the Receiver. Whatever the arguments it was now no longer relevant and the job of integrating this new initiative fell to me. Some baptism of fire!

At least the location of this new outpost was congenial. The farms were about fifteen miles from Newcastle upon Tyne and within spitting distance of the Roman Wall. It was wonderful open Northumbrian countryside and whenever I visited the area I had a sense of homecoming.

The first priority was to meet the staff, visit the farms and try to get my head round the whole project. Runge was keen that we should absorb all the staff within the MMB umbrella and persuade them that being part of a much bigger organisation was in everyone's best interest. David Storey had been the top man and his attitude to being subordinate to me was a crucial factor from the start.

David had an accountancy background and had worked his way up to become Premier Breeders' Chief Executive. He was tall, articulate, quick-witted and efficient. If you didn't know, you might have pigeonholed him as a city financier, not a man running a cattle-breeding company. First impressions are important and looking back mine were mixed. He was polite and courteous, but not forthcoming. The vibes were – you may be the boss but this is my show and don't interfere. He knew that his knowledge of the cattle breeding world was superior to mine and no doubt wondered why on earth I had been appointed to become Managing Director. For my part, I needed him to help me get to grips with the job of integration and the commitment of his former staff to the wider cause. It was to become a tortuous relationship.

Building a new image

Apart from the challenge of the newly acquired Premier Breeders' business, there were other urgent matters to attend to as soon as I had divested myself of the day-to-day running of the Farm Management Service. First, there was a need to fill the vacancy I had created. Tom Kelly was the man appointed. He had replaced John Butler some years previously when John had decided to leave the MMB and set up his own business manufacturing concrete cattle troughs and a wide range of other products appropriate to the technical skills of his new business partner. Tom was a Bangor graduate and had decided to change direction from a career in the Revenue to join us as a farm management consultant. His tax knowledge was a major benefit he brought with him. He lost no time in making an impression. Forthright, practical and fierce determination were his hallmarks and having served a number of years leading the North-West Region and Wales team, he was the natural successor to take over the leadership of my former department.

The other man I was intent on persuading to join my senior team was Steve Amies. Having served his time as a consultant he had moved to Head Office to work as PA to the MMB Managing Director, or by now the Chief Executive, Charles Runge. Steve took little persuading and I created a new position for him within the AI Service on the breeding programme side. He had been involved in the Premier Breeders' negotiations and his inclusion alongside Dick Shaw and James Booth injected a much-needed dose of financial realism.

Runge was anxious that we renamed and rebranded the Farm Services Division and recommended we retain a professional market research company to put forward some proposals. We had endless meetings with their staff and eventually they came up with the name of 'Genus' – pronounced as in Venus. The dictionary definition is 'of many species and being one of a series constituting a family'. It was a short word, with a scientific connotation and it didn't need translating into different languages. There was also an association with our main business which was animal genetics. It fitted the bill perfectly and remains the name of the company to this day.

The transformation involved a great deal of work such as logos, redesign of all stationery in the corporate image and a proposed restructure of the original departments into Genus Breeding, Genus Management (Records and Consultancy) and Genus Animal Health. Each head was given the title of Director, although none of us were independent of the main MMB machine. The proposals were well received by the main Board and following a vigorous media campaign we signalled that we had embarked on a new era under new management. Time would tell whether all this was simply a publicity stunt or was a real turning point in the fortunes of the old Breeding and Production Division.

The farm and goodbye to Mac Cooper

It became impossible to devote any consistent time to the farm during the week. We decided to put the whole acreage into grass and let the grazing on an annual basis, initially for cattle but within a few years for sheep. I had a number of tenants, mostly good but one disaster from North Wales. Notable amongst these was an old friend, Roy Willis, who ran a superb dairy enterprise only a couple of miles away. He reared some of his young stock at Spy Hill. Following that, a pedigree sheep breeder, Eifion Ellis from Pentrefoelas in North Wales had the farm for about seven or eight years, but eventually found the distance to be too time-consuming. Even after retiring at the end of 1996 I didn't relish the idea of starting to farm again, although I maintained the fences, sprayed thistles and generally kept the place tidy. I hate to see farms falling into disrepair.

In September 1989, Barbara's father, Mac Cooper died, aged seventy-nine. He had been unwell for some years with emphysema. In former years he had been quite a heavy smoker. As described earlier in this memoir, Mac Cooper had been my professor when a student and also we had a mutual love of rugby football. Having married into his family, our relationship matured and without doubt he was a major influence in my life, as indeed he was in many other people's lives. He was a gifted communicator both orally and with the written word. He was genuinely interested in young people and provided they worked hard, he would do everything in his power to help them along. Although a third generation New Zealander, he decided to stay in the UK after he retired and fortunately we managed to see a lot of both Mac and Hilary, his wife, during the latter part of his life. In 1986 they had celebrated their golden wedding with all their family at Spy Hill Farm – it had been a memorable weekend. He was an inspirational man and I will always be indebted to him for the help and guidance he gave me over thirty years.

CHAPTER SEVEN

The Early Nineties

THE CONSERVATIVE PARTY continued in power until 1997 when Tony Blair won a landslide election to bring in a New Labour Government. Mrs Thatcher's reign came to an end in 1990 following the introduction of the Poll Tax and increasing unrest throughout the country. She gradually lost the loyalty of her Cabinet colleagues and was eventually persuaded to resign. John Major took over as Prime Minister and although he won a twenty-one seat majority in the 1992 election, the economic stability of the country was soon rocked by the events of 'black Wednesday' with a run on the pound and the decision to suspend Britain's membership of the ERM. The Tories struggled on for another five years, then disappeared into the wilderness to lick their wounds and rethink their strategy, much as Labour had done.

On the international scene it was an eventful decade. The Channel Tunnel connected Britain to the Continent; the USSR disintegrated and the Cold War finally ended; wars in the Gulf and the Balkans, as well as continuous bomb outrages carried out by the IRA, ensured that our armed forces continued at the front line.

The Windsor Castle fire and the high profile Royal divorce followed in 1997 by the death, in a car crash, of the 'People's Princess' certainly made it a torrid time for the Royal Family. There were serious questions being asked about the future of the monarchy and much to their credit they seem to have re-invented themselves as a more acceptable institution.

As always some good things also happened. Apartheid ended and Nelson Mandela became the South African President. They even won the World Rugby Cup to celebrate. Real progress was made in Northern Ireland with the signing of the Downing Street Agreement and inflation fell below 2% and stayed low for the rest of the decade. Finally we made the eminently sensible decision to stay out of the euro when it was launched in 1999.

BSE

Large-scale outbreaks of foot and mouth disease, although catastrophic for the farmers whose animals are infected do not elicit a widespread panic within the general public, principally as it is well known that the disease is not transmissible to man. However, when a food scare breaks the consequences can be altogether different. Edwina Currie's famous remark in 1988 when she told ITN: 'Most of the egg production in this country sadly is now infected with salmonella' resulted in a 60% fall in egg sales overnight. Loss of income forced farmers to slaughter 4 million hens and destroy 400 million eggs. Many producers went broke.

BSE or bovine spongiform encephalopathy developed into another major food scare during the nineties. More commonly known as mad-cow disease it is a fatal neurodegenerative disease in cattle that affects the brain and spinal cord. The cause of the disease is widely thought to have been due to the relaxation of the rules governing the heat treatment of recycled material from dead animals to produce meat and bonemeal, a protein-rich supplement used in feed compounding. The real scare was that infected animals had been entering the food chain and a new human disease was identified and directly linked to BSE. It was known as 'new variant Creutzfeldt-Jacob disease' or vCJD. As the media frenzy took hold, wild claims were made that this disease would eventually cause thousands of deaths in people who had eaten meat from infected animals. In the event less than 200 people, to date, have died, but that is no consolation to the families affected. The effect on British dairy and beef farmers was dramatic. The EU imposed a ban on beef exports from the UK which lasted ten years and over 4 million cattle were slaughtered during the eradication programme. The damage done to the reputation of the farming industry was deep and long lasting.

Genus farm management consultants were extremely busy throughout the nineties trying to help clients whose herds had been affected by this awful disease and whose livelihoods were under continual threat.

Relocation to Crewe

Although the decision, in principle, to move away from Thames Ditton had been taken around the time of my appointment as MD, I only announced it publicly in January 1990. It came as an unpleasant shock to most of the staff based at Head Office that we were to relocate to Crewe. At that time it was not known that the MMB was to be dismantled and the Head Office building eventually demolished to make way for a new housing estate.

There were about fifty people involved and they had to make the difficult choice as to whether to move north or take redundancy. Furthermore Crewe, being renowned as a railway junction, seemed a big comedown compared with the leafy suburbs of Thames Ditton. I'm sure most people thought that I had engineered the whole thing so that I didn't have to move. It certainly looked that way and whatever reasons I gave fell on deaf ears. Most of the senior staff decided to move together with about fifteen others. Fortunately most of them were pleasantly surprised with the surrounding countryside and the financial advantage of selling a house in the south and buying one further north. We had leased a brand new building on a recently developed business park on the outskirts of the town and planned to move in during the late summer of 1990.

One of the reasons for choosing the location was the available labour market as we needed to recruit about thirty staff to replace the ones who had left the company. We also needed to engage computer specialists to run a mainframe system for billing and accounting purposes as our access to the MMB computer system would eventually come to an end. There was an excellent response to our job recruitment campaign and we had hundreds of applications from which to select good calibre people. I found an exceptional secretary in this campaign and Julie Bachelor (later Julie Simon) provided all I could wish for in professional support for the rest of my time with the company. It was also a golden opportunity to establish a new culture and build up a team work ethic.

Business under pressure

Although the excitement of creating a new image and the anticipation of soon becoming a separate company from the MMB was real, there was also the growing realisation that our main business was in a parlous state.

Dairy farmers were continuing to leave the industry. It was true that herds were getting bigger but the total number of cows was reducing year on year by between 2 and 3%. This inevitably meant a lower number of inseminations and semen sales. AI was the backbone of the Genus business and we had a 75% market share – the impact was unavoidable and cumulative.

Many farmers had already switched from the technician service to DIY, carrying out the job themselves. There was plenty of training available and provided the operator was competent, there was no telling evidence that conception rates were significantly lower in DIY herds. In fact advocates would claim that in many cases it was better because they could observe the animal in heat and time the insemination more accurately rather than having to await the arrival of the AI man. The switch to DIY was another reason for declining customers. Whilst we might still sell them semen, the opportunity to do so through the AI technician when on the farm was lost.

Reference has already been made to the fact that our policy for breeding new bulls for the AI stud had been misguided for many years. Perhaps it is easy to criticise with hindsight, but the fact remained that demand for home bred bulls was falling rapidly and British dairy farmers wanted high production bulls. And they wanted them now – not in five years' time when a change in direction might start to come through. We could obtain the semen they wanted from overseas and charge them an additional amount for handling it, but it was a delivery service not the result of a successful breeding programme.

The acquisition of Premier Breeders was not going to change things overnight. It might not change things at all. What was certain was that the culture of the two businesses was entirely different and it was to be a major challenge to integrate them under one banner.

The running cost of the AI organisation was excessive and was becoming more so as demand declined. There were around twenty main centres, most of which had sub-centres, scattered across England and Wales. All were staffed and managed by administrators and clerical employees. Many of the main centres also had bulls on site and extensive traditional buildings for winter housing and semen collection as well as laboratory facilities. A number even had adjoining small farms used for bull tethering and haymaking for winter feed. Bulls are dangerous animals to manage and labour costs were high in order to meet safety requirements. There had been no re-investment of AI profits when times had been good and the consequences when turnover fell were self-evident. It was obvious that the whole organisation had to be completely restructured, both to massively reduce costs and at the same time provide a more efficient service to a smaller total market.

Breed society politics

Politics always added to the pressure on the business. Originally for animal health reasons, the Ministry of Agriculture controlled who could operate an AI service under licence. The MMB had the lion's share (75% in England and Wales), with two or three other much smaller co-operatives operating in the Midlands and the south-west. It was becoming increasingly likely that the Ministry were going to abolish these monopolies so that anyone could set up and provide an AI service.

The MMB and now Genus still analysed and published all the information concerning the test results of dairy bulls (both home bred and overseas) used in the UK. As the competition for semen sales accelerated so did the demand that the control and publication of these results should be undertaken by an independent group, especially as Genus was now assuming a more commercial role. The MMB had recently commissioned the Wilson Committee to make some proposals to this effect and the outcome was that an independent Animal Data Centre was to be created to take responsibility for this work. The main cattle breed society had been lobbying to get its hands on the Animal Data Centre and was also anxious to take over National Milk Records.

The Holstein Friesian Society argued that they had the experience of dealing with many of the same farmers through their pedigree registration procedures and also the field staff infrastructure to carry out daughter inspections for type classification. In their view there could be significant synergy and cost saving for the industry in amalgamating the two organisations. Not everyone agreed – we certainly didn't.

In the event NMR eventually became a company in its own right with a Board and Chief Executive. Tom Kelly had earlier persuaded the MMB that it had to remove the 30% subsidy, so at least it became a viable entity before the MMB was abolished. It would never be a commercial success in terms of returns to shareholders but it could reasonably be expected to break even and continue to provide farmers with essential management data as well as sell its collective records for bull testing purposes.

My first direct encounter with the politics surrounding the Holstein Friesian Society was at the Cattle Breeders Conference in Cambridge. It came as a bit of a shock. All the influential breeders in the industry get together for this annual event and whilst the gathering had some similarities with the British Society of Animal Science, it was much more politically orientated as well as being dominated by practising dairy farmers rather than research scientists. Central to the debates were the views of the breed societies.

The Chief Executive of HFS was a man called Duncan Spring. He was a qualified accountant and a skilled administrator. In a paper to the 1989 conference, no doubt encouraged by his Board, he attacked the MMB cattle-breeding record and roundly criticised the 'arrogance' of those in charge. He went on to say that whilst thirty years previously Britain had led the world in this field, now we were reliant on imported semen because the demand for British-bred dairy bulls was in freefall. It was all a bit unnerving to say the least and sitting next to Kevin O'Connor, I expected him to mount a vigorous defence. Sadly there was little ammunition to hand. Spring even alluded, I felt rather sarcastically, to a new era about to dawn at the MMB as they had recently re-invented their Breeding and Production Division as Genus – 'whatever that means', he intoned.

It was an uncomfortable few days. I knew many of the delegates and most of them offered their congratulations and good wishes in my new role. But I could also see the doubt in many of their faces when they wondered how someone with a recognised farm management reputation had suddenly found himself in the hot seat of AI and dairy cow genetics.

Another important dimension in this debate began to emerge shortly after this conference and perhaps stimulated by it. Tim Heywood, the Managing Director of the Duke of Westminster's Grosvenor Farms had persuaded the Duke that he should invest serious money in establishing a new bull-testing programme on his estate in Cheshire. This initiative became known as Cogent and under Tim's capable leadership and his fierce determination, it materialised into a significant force for change over the next five or six years. Such is the length of time before any results can be produced from a bull-proving programme.

I attended the Cambridge Cattle Breeders Conference regularly until I retired at the end of 1996 and things gradually improved as we signalled that we were intent on doing something about the problems. Although there continued to be much politicking between HFS and Genus, especially with regard to who was to run milk recording and where the Animal Data Centre was to be located. Duncan Spring and I eventually (after we retired) became good friends, despite our many disagreements at that time.

So what were the business priorities?

The main problem with our own breeding strategy for the last decade or more, had been the lack of vision with regard to changing demand for dairy bull semen. Original policy had been to improve both the milk and beef potential of the black and white cow. Farmers now wanted bulls which had milk-yield potential far beyond anything we might reasonably expect to achieve from our programmes for at least the next six years. We had over 600 unproven bulls in the production line from calves of a few months old to six-year-old animals awaiting their progeny test results. When these were published the decision would be made to bring them into the national stud and market their semen or have them killed for beef. The stark fact was that none of them was ever going to be good enough because the international competition was so superior. We had to destroy them all and start again. The MMB Board were understandably appalled at this admission of failure which would not only result in a heavy financial loss but also give the media and our competition a field day. The problem had been there for years and it was Steve Amies who convinced us all that the nettle had to be grasped. After many torrid meetings the MMB Board eventually agreed to the night of the long knives and all the bulls were slaughtered. It was a tough call but had we not made it then, I don't think Genus would have survived those next few critical years.

The other immediate problem concerned the purchase of dairy semen from our overseas suppliers, principally from the US and Holland. Demand for the top international bulls seemed to have no ceiling. Prices for semen from the top bulls went through the roof. In one case we held an auction to sell the remaining supply from a bull named Blackstar and prices reached £200 for a single dose. This was obviously exceptional and interest was confined to pedigree bull breeders but the market was buoyant.

Our policy had been to identify those bulls which were considered likely to be in greatest demand and estimate the quantity of semen required to meet that demand. It was a complicated process because of import restrictions which required orders to be placed many months in advance. The problem was that the staff who had been doing the buying had little understanding of the financial consequences of inventory management. Demand was continually changing, one super bull one month, another the next as the semen-selling businesses chased market share. Naturally we ensured that we had an ample supply of a popular bull but when it went out of fashion, a high proportion of the semen became unsaleable – not unlike the sub-prime market which triggered the last banking crisis. Whilst we had made a reasonable return on the difference between the selling price and the purchase price, the loss on the unsold stock was total. Again Steve Amies discovered this black hole as he delved deeper into the accounting records of the AI business. There was no escaping the effect that this had on our accounts as years of accumulated inventory losses had to be cleansed from the system. The reputation of Genus, although it had started so well, was in rapid decline. The chickens had assuredly come home to roost and I had to absorb the mounting anger of the MMB Board at their monthly meetings. It was a stressful time.

Recovery from a low point

As Christmas approached at the end of 1990, I began to realise I was in trouble. I slept very little and was continually exhausted. I couldn't get my mind away from the awful financial results which were accumulating and our seeming inability to do anything quickly to avert disaster. I wasn't able to concentrate and would start to read something for about ten minutes and realise I hadn't taken in a single word. My mind was out of control. I talked to Charles Runge who tried to put things in a wider perspective. I even met with the Chairman of the Board, who was sympathetic, but no one could wave a magic wand. Charles insisted that I took a couple of weeks' leave and recommended long walks with the dog. But I was consumed with self-pity and dreaded even going out socially as I thought everyone would think I was having a nervous breakdown. It almost became a self-fulfilling prophesy and my friends did begin to notice that something was seriously wrong.

Barbara patiently encouraged me to see a specialist and although I was prescribed something to make me sleep, there seemed no way forward through the problems I perceived. She even hid the keys to the gun cupboard. I did come out the other side thanks to Barbara's careful encouragement and loving care, but it was a close run thing. When I'm now told someone is seriously depressed, I have an inkling of what they are going through. It was a salutary lesson.

Early in the New Year I struggled back to work. My friends and colleagues were hugely supportive and I gradually began to regain confidence. The incident, however, wasn't forgotten. I was sure Runge felt that I had failed the test and he should be looking for a new CEO. He had immense problems of his own as the MMB Board began to face up to its own dissolution and what was to happen next. He couldn't afford to carry a lame duck. I believe he had David Storey in mind to take over from me. They had worked together on the former Premier Breeders takeover and I knew he had a high regard for David. Certainly they began to discuss strategic issues outside our Directors meetings. I was being sidelined. Later in the year Charles came up to Crewe for the usual monthly Directors meeting and completely lost his temper with me in my private office. I let him rant and swear away without offering any defence. It was clear he wanted me to throw in the towel. Fortunately by then I was on the mend and determined to re-establish my credibility.

Suddenly Charles Runge left the MMB. I think the main reason concerned the arguments then raging about the future of Dairy Crest, the separate manufacturing company owned by the MMB but managed by its own Board. Charles wanted to influence the way this was to be positioned in the future and he was consistently denied access to the discussions. I suspect he lost his temper and walked out – alternatively he may even have been pushed.

George Wright, who had been the MMB Finance Director for many years was elevated to Chief Executive, and now became my new boss. An experienced accountant, he wasn't going to set the world on fire but he was undoubtedly a 'safe pair of hands'. It was to be an interim appointment whilst a search commenced for a successor to steer the MMB towards what everyone by then had come to refer to as the 'new world'. George had limited knowledge of the Genus business and seemed quite happy for me to get on with it. It also cleared the decks and I knew this was the opportunity to deal decisively with David Storey. Our relationship had never been a productive one. He really had no meaningful portfolio, having relinquished the responsibility of restructuring the Marketing Department after my insistence that he could not appoint one of his former colleagues to head it up. Neither was he prepared to move to Crewe to take over from Dick Shaw when he took early retirement. It was a difficult final meeting in Northumberland when, with a personnel man in tow, I informed David of his redundancy. He left the company the same afternoon. We had never got on and I sensed a palpable sigh of relief from the other directors when the deed had been done.

Genus restructure

Dick Shaw had been employed as one of the AI Veterinary Surgeons, first at Ruthin in North Wales and then at Head Office, eventually taking over as head of AI on John Frappell's retirement. I believe he had hopes of succeeding Kevin O'Connor, but having seen the problems I had to cope with I suspect he was relieved he didn't get the job. Dick had been a loyal colleague and had worked hard to help the organisation succeed when he moved north to Crewe. He struggled, however, with the financial side of things and could see his influence in managing the AI business beginning to decline as Steve Amies got to grips with the problems and possible solutions which would lead to significant redundancies in field staff. Dick therefore decided to take early retirement.

He was replaced by Bob Weir who joined the business without any background in farming matters, let alone AI. He was an exceptionally bright individual, a computer expert and an able communicator. My thinking was that we already had plenty of technical expertise within the business, what we needed was the brains and vision to restructure the whole field operation and at the same time take advantage of the rapidly developing data collection opportunities which were coming onto the market. I knew his appointment was a risk but hadn't appreciated how highly strung he turned out to be, nor the fact when under pressure, he drank too much and made a fool of himself. He stayed on until 1997, but then left and sadly died at an early age.

The restructuring of the AI business preoccupied our thoughts and actions for months. Many of the main centres were amalgamated, most of the sub-centres were closed and the bulls re-housed in the more labour-saving units. There was a substantial redundancy cost, but most of that could be met from the eventual sale of the assets which were no longer required.

It was a painful process having to travel up and down the country to tell the staff, many of whom had been employed by the MMB for years, that they were to lose their jobs. In addition, we had to meet with all the Regional AI Committees to explain that this layer of farmer involvement was to end and that the company was now set on a commercial course and simply couldn't afford the time to continue this activity as well as fund the cost.

The end of the MMB

Andrew Dare became the new and final Chief Executive of the MMB in the late summer of 1992 with the task of leading the dairy industry into the 'new world'. Andrew had been in the manufacturing side of the business for most of his career. He had been working in Ireland for Unigate Creameries and had been persuaded to act as a consultant to the main MMB Board. They liked what they saw and he was duly appointed to the top job. He was a tough cookie. He formed an Executive Committee which met every Monday morning at Thames Ditton at 10.00 a.m. The late start was a concession to allow me to get there in time without having to stay away every Sunday night. He rarely gave concessions and soon started to weed out what he considered to be the weak links in the senior management at Head Office.

There was a massive government review in progress as to what was to replace the MMB when it was finally abolished. It was clear that it would lose its statutory powers and farmers would be able to sell their milk to whomever they chose. There was nothing to stop farmers continuing to run a co-operative to market their milk, but it had to be voluntary not compulsory. The big dairy companies and co-operatives were already busy courting the larger producers with promises of eye-watering prices for supply contracts. But for the smaller dairy farmers the prospects looked less promising. The MMB plan was to launch a successor company to be called Milk Marque. Andrew and his staff had the task of convincing farmers that it was in their best interest to join up. At countless meetings throughout the country he urged them, especially the bigger ones, not to fall for inflated prices from buyers. It was essential they stuck together and retained the power of collective bargaining. Milk Marque succeeded in capturing about two-thirds of the supply from the now 30,000 producers, but this wasn't enough milk and the new co-operative was left with too high a proportion of the smaller more remote farms resulting in higher transport costs and less volume over which to spread overheads. What happened was exactly what Andrew had warned against. Once Milk Marque had been forced out of business, the independent buyers were able to pick and choose their suppliers and force down their prices. Towards the end of the nineties, dairy farmers were receiving almost 20% less for their milk than they had been before the MMB was abolished. It was the final nail in the coffin for many of the smaller more isolated dairy farms.

At the same time as the MMB approached its dissolution, we had to make plans for Genus to become a free-standing company. Vesting Day was to be 1 November 1994 – and after that date we were on our own. There was a great deal to do.

CHAPTER EIGHT

The Rest of the Nineties

Genus becomes a limited company

GENUS AND ITS ASSETS were to be distributed to almost 30,000 registered milk producers in the form of shares based on their previous year's milk production. Rather like quotas, this new asset would appear out of the blue, but this time the perceived value would be small, and for the average farmer would only amount to a nominal value of a few hundred pounds. Until a share trading mechanism became available, it was anyone's guess what the real value of the shares might be. Most of the share certificates simply disappeared behind the clock on the mantelpiece. A significant number of recipients had never before owned a share, never cashed a dividend and have now disappeared without trace.

Andrew Dare was preoccupied with the MMB wind down and delegated to me the job of working with the lawyers and the merchant banks to prepare the ground for Genus to become an independent limited company. Freshfields seemed to me exceedingly tedious as the legal requirements necessitated dealing with reams of paper to meet the laws of 'due diligence', the formulation of Articles of Association and much inexplicable detail which only lawyers can understand. Goodness knows what it cost – fortunately the MMB were picking up the bill. I couldn't even write a letter concerning our progress to outside organisations unless it was approved by the lawyers.

Working with Lazards, the merchant bank, was much more interesting. Nigel Turner was the main contact and I spent many hours in their offices in London. The most immediate priority was to get the MMB to agree to the appointment of a Chairman of the new Genus Board. I was determined that we did not inherit any of the MMB Board members, many of whom had previously been involved with Genus in an advisory capacity. Aware that the MMB was nearing its end, some were actively lobbying for inclusion and even the position of Chairman. I already had someone in mind for this key post. Once he was accepted by the MMB we could begin to appoint our future colleagues. Nigel Turner's help in this was crucial. If the proposal came from me, I felt sure it would be rejected. If it came from Nigel there was a much greater chance of success – he was independent and had no axe to grind.

I approached John Beckett at the Royal Show in July 1994 to see whether he might be interested in joining the new Genus Board and even to consider becoming Non-Executive Chairman. John was and indeed is, a remarkable man. A gifted athlete in his younger days he was struck down with polio at the age of eighteen. He has worn a brace on his leg ever since. He has never allowed this to limit his activity or intentions. He might look a bit unusual swinging his useless leg along with him, but with the aid of a stout stick he gets there a great deal quicker than you might expect. He had developed an impressive dairy farming business near Whitchurch in Shropshire and also manufactured cheese. Known as Belton Farms, the reputation of the company went before it and he and his staff have been the recipients of many awards for excellence. I had visited the farm on numerous occasions in the course of my farm management work over twenty years previously and I had a great respect for John's determination, judgement and integrity. We had become good friends.

He also, crucially, had an impressive commercial track record. Amongst other things he had been Chairman of North Western Farmers for eleven years. NWF was a successful co-operative supplying customers with compound feeds, fuels and much besides throughout the north-west of the country. It was a well-managed and expanding business with an enviable reputation in a highly competitive field.

After considerable thought he agreed to allow his name to go forward as prospective Chairman of the new company. Nigel Turner had little difficulty persuading the MMB that we had found the right man and they approved the appointment. The next priority was to identify and interview the remaining non-executive directors. It was probably unusual for the Chief Executive to sit in on this process, but I was the only one who had detailed knowledge of the business and also some of the candidates. We saw quite a few people including some former MMB members, only one of whom, Cedric Scroggs, was successful. He had been a Government-appointed member of the MMB, rather than an elected farmer and he brought with him some valuable business acumen which he had accumulated whilst formerly working in the City. I felt it important to include other well known farmers as our shareholders still perceived us to be primarily a farm service provider. Two other dairy farmers were thus appointed: Guy Trehane from Dorset, whose father Sir Richard, had been a former Chairman of the MMB and Edwin White from Somerset. Both had business experience outside farming and were well known political operators. Tom Neville, a chartered accountant and former Finance Director for Rolls Royce completed the non-executive team of directors.

On the executive side, besides myself there was Steve Amies and Bob Weir responsible for the breeding business and Tom Kelly who ran the farm management side. In addition we had recruited a new Finance Director, Philip Acton, who was an accountant and had previously been FD for a public company. David Fairall, head of our Personnel Department, became Secretary to the Board so we became a total team of eleven people.

Computer problems

A major advantage of being part of the old MMB had been the ease with which we got paid for the goods and services provided to our customers, something we had taken for granted for years. Dairy farmers also considered the regular payment of their entire previous month's milk supply to be one of the biggest benefits of the MMB. The monthly milk cheque netted off the costs of any services provided and an automatic transfer of funds was then credited to the Genus account. There was no need to chase up late payers and it also rather hid from view the detailed list of expenditure items. It was a painless exercise and one we were going to sorely miss when we had to invoice and collect our own fees when the MMB disappeared. We knew we had to build our own system and had been working on this ever since moving to Crewe. We didn't know, however, what a mess we were going to make of it.

Almost every day one reads of the abandonment of some vastly expensive mainframe computer system as the costs escalate and the output fails to provide the users with the information they need or were led to expect. A blame culture soon develops with the contractors complaining that the users keep changing their minds. Many senior staff lacked basic technical understanding of computer systems and had to rely on the specialists for guidance – I know I fell into this category.

We had negotiated a contract with one of the main software agencies designed to allow us a few months of dual running with the existing system so that the bugs could be dealt with before the MMB closed down at the end of 1994. In the event deadlines were missed and promises broken so that we went live with very little testing resulting in an inevitable shambles as hundreds of customers received the wrong invoices. Worse was the fact that they were expected to pay for what they had received, rather than have the money deducted from their milk cheque. Naturally where there were errors or queries we didn't get paid until they were sorted out to the farmer's satisfaction. The effect on our cash flow was serious and the time to resolve the problems was cumulative. Farmers are known to have little patience with administrators and our front-line staff came under intense pressure and criticism through no fault of their own. Many had to endure abusive phone calls and morale started to fall sharply.

The problem took months to sort out. We lost income as some customers refused to pay, others we had to write off as bad debts. Equally galling was the adverse publicity that the whole debacle stirred up and our new Non-Executive Directors, initially understanding, soon began to wonder what they had let themselves in for. We eventually sorted out the problems but it was a dispiriting experience and an unfortunate start to an independent relationship with our customers. Nevertheless we managed eventually to persuade about 60% of our customers to sign up for direct debit payment within the first year – a remarkable show of loyalty after such a bad beginning.

Share trading

The scheme of reorganisation required the new company to make reasonable efforts to arrange a facility on a periodic basis for the trading of shares in Genus Limited. As a first step we arranged for Lloyds Bank Registrars to set up two specific dealing days during the first year, whereby shareholders could offer to either buy or sell shares at specific prices. These offers were then matched and prices falling within the acceptable ranges were exchanged. Only a very small number of shares changed hands (less than 2% of those in issue) with a clearing price of 47p and 51p on the respective days. So now we had some idea of the value of the shares, but as sale volumes were so low, the indication was simply a best guess.

It was to be some months before interest in Genus shares began to increase. We needed a more visible and regular market and the Alternative Investment Market or AIM seemed to fit the bill. This is designed for small emerging companies and doesn't require the participants to meet the demands of a full listing on the Stock Exchange – this was to come later after Genus became a public limited company. One of the problems was the vast numbers of small share allocations in circulation, and many of the recipients had simply forgotten they had them. We really needed to interest some institutional investors if we were to take the market seriously and to do this we incentivised small shareholders to cash in their holdings to increase purchase opportunities. It was also important that we encouraged staff to invest in the business and a Savings Related Share Option Scheme as well as an Executive Share Option Scheme were also approved at the 1997 AGM. Unfortunately the share option scheme didn't become functional until after I retired but I did buy a few thousand shares on the original matched bargain basis. They turned out to be an outstanding investment.

Independent Genus becomes established

Despite the many challenges faced since Vesting Day, we had set the foundations for a successful business with a first-year operating profit of £827,000. The Chairman in his annual report for the year ending March 1996 stated:

> Although this is the second Genus report to be published since the company was launched in November 1994, it is the first to include a full year's set of accounts. Whilst the problems of transferring from an MMB division to an independent company are now largely behind us, the changes taking place, especially in the breeding industry, continue to present major challenges as competitors intensify their efforts to maintain market share. Such activities benefit customers as competition provides choice and only by providing value will suppliers improve their position. Great emphasis is placed on this objective within Genus and most customers using the technician service AI Contract would agree that, not only have service levels and genetic quality of semen improved, but prices have also fallen significantly over the past two years.

In view of the tough trading conditions, a market in overall decline, computer problems and the difficulties caused by BSE, at least it was a positive start. Our borrowings were negligible and the expectations high that once the restructuring of the AI business had finally settled down, there was every chance of building on these results the following year. We had also launched new products and initiatives aimed at increasing income and more fully utilising the sales opportunities for our field staff when on farms.

By the end of the following financial year, my last with the company, operating profit had risen to £1.9 million on a turnover of £47 million. The Board even decided to pay a modest dividend of 3p per share – not a lot, but a step in the right direction. Encouraging progress had been made in most areas and it was clear we were not going to sink without trace as some commentators had predicted.

Since then I have been a keen observer of company growth and influence. The acquisition of a major American cattle-breeding company (ABS) in 1999, followed by that of the world's largest pig-breeding company (Sygen – originally the Pig Improvement Company) in 2005 have been the major strategic moves. It is probable that these two acquisitions have transformed Genus into the biggest animal genetics company in the world, trading in over seventy countries, including the Americas, Russia, China and India and firmly establishing the 'Genus brand' on a global basis. By 2010, annual sales revenue had reached £285 million, generating an operating profit of over £39 million

22. Our wedding day, 1962. My father and Lillian standing on the left.
Mac and Hilary Cooper on the right

23. Figures and farming, 1965

24. Farm Management Consultant,
Cheshire, 1966

25. Spy Hill Farm, 1965

26. Spy Hill Farm, 1975

27. With Sue and Pete above the farm, 1966

28. Contract rearing dairy heifers, 1966

29. Harvesting barley with a broken arm, 1966

30. Consulting Officers Conference, 1977

31. With John Frappell on a consultancy visit to India, 1978

32. Family Christmas at Spy Hill with the Thompsons, Morrises, Coopers and Cravens, 1973

33. Family camping holiday in Ireland, 1976

34. Mac and Hilary's Golden Wedding celebrations at Spy Hill Farm, 1987
From the left standing Neville Sorrentino, John Craven, Robin Craven, Peter Craven, Sally Morris, David Morris, Fiona Thompson, Phil Thompson, David Craven, Dai Morris Sitting: Val Craven, Barbara Craven, Hilary Cooper, Sue Sorrentino, Mac Cooper, Cynthia Morris, Diana Thompson

35. My father and step-mother on their wedding day, 1965

36. Hilary and Mac with their three daughters, Diana, Barbara and Cynthia, 1987

37. Family holiday in the British Virgin Islands, 1999

38. Skiing in Zermatt
From the left: John Willis, John Craven, Pat Willis, Barbara Craven,
Alison Tresias, and Richard Tresias

39. Presenting the Genus-sponsored Perpetual Trophy to Murray Stevenson, President of the Ayrshire Cattle Society, Dairy Farming event, 1966

40. Retirement gift presented by John Beckett, Chairman of Genus, with my secretary Julie Simon ensuring safe delivery of the tray and decanters, 1996

and a dividend of 12p per share. By mid 2011 the share price was in the region of £10 – a far cry from the 50p realised when the shares were first traded fifteen years previously. By any standards it is a remarkable success story, built I like to think, on the solid base established when it emerged from the wreckage of the MMB. From where I sit, Genus looks set to go from strength to strength. It is now a leading player in the increasingly high-tech world of genetics, a world where size is crucial for research to pay for ongoing technological development. It also operates in a market which seems sure to grow as world population and demand for food shows no sign of slowing down.

Time has moved on. No farmers now sit on the Board – they are all high-powered City professionals, no doubt intent on producing the best possible returns for their shareholders. I suppose there is always the risk that the company might some day become the subject of a takeover bid, even a hostile one, perhaps by a predator interested in stripping out the valuable assets and splitting up the business. There are now almost 60 million shares in issue with farmers collectively only holding about 15-20%. So if a bid did materialise, it will be the major institutional shareholders who would determine whether or not it was successful. Also there might be less scope than expected for asset stripping as a substantial part of the real value of the company seems to be tied up in 'good will'. Whatever happens, I hope farmers remain significant shareholders, if only because they also represent the customer and their opinion ought to matter to those who decide future strategic direction.

Succession

I was due to retire on my sixtieth birthday in August 1997. The search for my successor commenced towards the end of 1996 as it was possible that the successful applicant might have to serve a six-month period of notice. The Board had appointed a Nomination Committee consisting of the Chairman, Cedric Scroggs, Guy Trehane and myself to interview the short-listed applicants and propose a new Chief Executive. In the final analysis it came down to two people – Steve Amies and Richard Wood, formerly the MD of a division of Lloyds Chemists plc and previous to that, MD of ICI Seeds (UK). I was convinced that Steve would do an excellent job. I had worked with him for many years and felt that his knowledge, enthusiasm for change and sheer intellectual ability made him the obvious choice. The others didn't agree. Richard, a slight, rather dapper man, came across well and advocated an entirely new approach away from the traditional farmer co-operative culture which both Steve and I had grown up with. The others felt Genus really had to change gear and embrace commercial reality – we were both considered 'farmers' men' and not sufficiently experienced in the ways of the City. Judging by the subsequent financial results, it seems they were right. Steve stayed on and continued to play a leading role in the expansion and development of the cattle-breeding sector, but it was Richard Wood who was to drive the company forward.

Richard was able to join Genus at the beginning of December and after a brief induction period we decided that the sooner he got stuck in the better. I was very happy to make way and duly resigned later in the month, although the company kept me on its books until the end of my contract the following August. On 19 December, our thirty-fourth wedding anniversary, the Board gave me a splendid retirement dinner at the Grosvenor Hotel in Chester.

Reflections

To work for one organisation for the whole of one's career is becoming the exception rather than the rule. One might well argue that a change of direction is a stimulus and challenge to better provide the ambitious individual with a wider experience and a better CV for the next job. I only applied for two jobs whilst working for the MMB. One was with the then Midland Bank, which had a powerful and well-paid agricultural team led by Bob Bruce, an uncompromising Scotsman. It was at a time of frustration when for a short period my salary was actually lower than some of my colleagues who were junior to me. I didn't get the job and in view of later events, was very pleased I didn't. Anyway, I don't think Bob and I would have gelled – we were too self-opinionated.

The other job was more interesting and very different. It was as the Principal of the Royal Agricultural College at Cirencester. The College had, probably unfairly, a reputation as the playground for the farming sons of well-off landowners. In reality it was a well-respected, independent, residential institution dating back to the mid-nineteenth century with many famous agriculturalists as members of its alumni. It had an imposing cloistered façade with courtyards and extensive sports fields on the outskirts of the town. There were also some 1,200 acres of farmland attached to the college. I knew the incumbent Principal at the time, Sir Emrys Jones, formerly head of the Government Agricultural Advisory and Development Service and he encouraged me to apply. At the shortlist interview, things seemed to go well until the realisation that we would be expected to live in the Principal's house on the campus. This was the early eighties and once again Spy Hill played a role in the decision-making process. Obviously had I got the job we would have had to sell up and move, but we wanted to continue to have a small farm and roll over the inevitable capital gains into another enterprise. Whether, had

I not raised the issue of living off site, this would have influenced the panel to decide to make an offer, I don't know. I suspect they thought that my private circumstances were an unwelcome deviation and my total dedication to the job might be compromised.

I have no complaints about working for the Milk Marketing Board for the whole of my career. At its core were elected farmer members. There was a matrix of Regional, AI and NMR Committee members – all in place to oil the wheels of communication up and down the chain. This may sound tedious and over-political and at times it was, but for the most part I enjoyed dealing with them. They were nice people and wanted their organisation to be successful. Also I had spent my early years visiting individual farmers on their farms and trying to help them become more efficient and more successful. They worked hard, often in difficult weather and sometimes under severe financial pressure. They had to be resilient as the goalposts were forever changing. They had to have a sense of humour otherwise they would have gone mad dealing with red tape and bureaucracy. I respected their commitment and dedication, not just to their business but for their responsibility to the environment and to their animals. I think I understood them.

Three things stand out for me when I look back over my career. First, I was allowed to live in and run a small farm as well as do my full-time job as a farm management consultant. There were probably half a dozen of us doing a similar thing and we were trusted not to take advantage of that concession. I'm sure none of us did. To be able to return home to the farm and family after a hectic week was a wonderful bonus. It was also remarkable that we managed to continue to live there for virtually the whole of my career.

Second, I was encouraged to become involved with outside organisations and as a direct result visited dozens of countries in the course of this work, sometimes undertaking specific overseas consultancy projects.

Third, I was lucky. I was in the right place at the right time when fundamental change overtook the MMB and I had the unique opportunity to help establish Genus as a freestanding company. I wasn't the man to take it to its new heights, but I got it into a position from which this could happen.

My main regret was the way in which the MMB disappeared. The consequence for so many small dairy farmers was terminal, especially in the more remote parts of the country. Losing their collective bargaining power was a critical factor and the original reason for establishing the organisation in the first place – once described as 'the sheet anchor of British farming'. But it also has to be said that there were many larger producers who were pleased to see the end of the monopoly powers. They had been complaining for years that it was unfair that they had to pay the MMB a levy in order to add value to their product by selling direct to the public or making cheese themselves. It was restrictive practice and constrained business initiative. They welcomed the arrival Mrs Thatcher's administration.

Whether, with a more inspired leadership, there may have been a workable alternative to total abolition is an open question and now of interest only to those who were around at the time – over seventeen years ago. There was talk of splitting the organisation into entirely separate regional co-operatives, each with a share of the Dairy Crest business. This would avoid the accusation of monopoly as they could have competed against each other and yet in some way retained the responsibility for providing a market for the smaller more remote farms.

Easing into retirement

I was keen to continue doing some work after retiring from a long career in the dairy farming industry. The easiest option would have been to explore possibilities in the same sector or even going back to consultancy work. But I wanted to do something completely different.

The medical world held some fascination for me. Barbara had been working as a GP for a number of years and I therefore knew a bit about the NHS and the changes which were taking place to try and improve both its effectiveness and accountability. In the nineties the Government was attempting to introduce a more commercial approach to GP practices which were then in the throes of what was termed 'fundholding'. Rather than simply refer patients who needed specialist treatment to hospital, they were obliged to work within a budget and negotiate treatment costs. It came as a bit of a shock and it meant that they had to try and run the practice on businesslike lines, but at least it focussed the mind on the commercial reality of primary care.

At the same time, services run by the local community, such as health visitors, dentistry, chiropody and school health programmes came under the microscope in an attempt to provide better value for money. I replied to an advert seeking non-executive directors to join the Chester and Halton Community Healthcare Trust as they seemed to be looking for people who had business experience and an interest in the future of the Health Service. Following interview I was appointed as one of five non-executives, some of whom were local political worthies and others who had a background in social services. In order to get some idea of how the system worked I spent some time out in the front line with nurses visiting patients in their homes in the deprived areas of Runcorn. It made me realise what an amazing job these staff did. Many patients were elderly and housebound, living in very basic conditions. Their spouses usually played the role of carer and the visits by the nurses were their lifeline. The Health Service comes in for a lot of criticism, often from people who know very little about it, but the dedication and professionalism I observed in those visits certainly left a lasting impression.

The Chairman of the Trust was Professor Ian Stanley, a scholarly academic at the University of Liverpool. It was not an onerous commitment, consisting of monthly Board meetings at which senior medical staff reported on their various activities. With my commercial experience I was expected to take a special interest in the accounts and ask leading questions of the Finance Director as to the meaning of the monthly figures which poured forth unremittingly from his office. NHS accounting practices take some understanding and whilst our aim seemed to be to balance the books rather than make a profit, there was little real scope for improving the numbers. Getting rid of surplus administration staff was a major obstacle. There always seemed to be a reason not to act or to redirect people into another vague job title rather than face the hurdles of unfair dismissal or redundancy. Discontinuation of uneconomic services was also virtually impossible, as patients would suffer, so that any measurable progress towards a more cost effective organisation remained a theoretical pipedream rather than a practical reality.

After about five years another restructure of the NHS arrived in the form of the creation of Primary and Secondary Health Care Trusts. The Community Trust I had been involved with was abolished and its activities redistributed amongst neighbouring organisations. I was invited to transfer to the West Cheshire Primary Health Care Trust. This had a much wider remit and a bigger budget. It included all the GP practices as well as other primary health care providers such as dentists, pharmacists, and a multitude of other specialists. One of our main jobs was to negotiate a contract with our secondary care colleagues in the hospitals, both at the Countess of Chester in the city and institutions further afield.

It was an interesting, if frustrating, ten years working for the NHS. The main problem seemed to be the constant change and reorganisation. No sooner had one system begun to bed in than along came another root and branch restructure. Armies of statisticians monitored numerous targets used to judge progress. I have no doubt much time had been wasted massaging the data so that it bypassed the category under investigation and lay gathering dust in some less obvious place.

It's easy to become cynical about the whole multi-billion pound operation and the way in which it features so prominently in almost every political debate. The problems and costs are bound to continue to escalate as the population ages and life expectancy extends into the eighties and beyond. Political parties, whatever their colour, shy away from the essential truth that people will have to pay more in future towards the cost of their health provision. But in the final analysis I cannot speak too highly of the professional care and the exemplary dedication of the vast majority of those working for the benefit of their patients, often in difficult circumstances.

Home Office Extended Interviews

I found out about the Extended Interview process from a newly appointed Assistant Chief Constable in Cheshire, Brian Baister. He was a keen rugby fan and golfer so our paths crossed shortly after he arrived in the county, having previously been working for the Met. in London. Discussing possible part-time jobs one day with him he explained the process by which policemen were selected to undertake the Senior Command Course and if successful, qualified for promotion to the top ranks of the force. Although the selection panels included Chief and Assistant Chief Constables, there was also a Non Service Member who brought an outside dimension to the process. He recommended I make contact with the relevant Home Office department in London. My face seemed to fit and after various induction courses I joined the pool of NSMs.

My first session, as an observer, was held at a prestigious venue – the Grand Hotel in Eastbourne over a six-day period. It might sound like a holiday jaunt, but in fact it was extremely exacting work for both the aspiring applicants and their interrogators. Each group of six candidates had to complete detailed written evaluations and recommendations in answer to a complex business problem, having studied it for an hour or so beforehand. This was followed by a series of sessions where they discussed a wide range of topical and service issues, both as group members and chairman. Formal presentations and individual personal interviews completed the two-day assessment. The panel marked the candidates immediately following each session based on a series of competencies such as analysis skills, communication (written and oral), judgement, intellect, self-confidence and leadership. The reasons behind each mark awarded had to be written up and eventually returned to the candidate after the results had been declared. There were five different panels operating at the same time and everyone came together at the end of the two and a half day process to put forward their deliberations and to recommend those whom they considered good enough. After a quick lunch the whole process started again with a new batch of candidates. By the end of the week most of us were mentally exhausted.

Only about 20% of candidates were successful, although those on the borderline were encouraged to have another go in the future. I was impressed with the format and whilst no interview process is infallible, this one must rank as one of the best I have come across. It was also used in the Fire Service and in the Prison Service and I was involved in a number of those interviews. The other benefit for me was to meet and work alongside senior people in these vital public services and to learn about the problems they faced when dealing with their political masters and local authorities. They were an impressive bunch.

I was even asked to put together a complex problem for one of the police sessions. I based it on the emergence of the Genus business – excluding any direct reference to the company or its staff. It sought to test the candidate's analytical skills and financial understanding as well as seeking his recommendations as to whether the right strategy for the future might be to focus on a farmer co-operative business or go all out for a commercial approach. Out there somewhere are probably a handful of Chief Constables who can remember their extended interview experience and having, at one stage, to apply their minds to maximising sale volumes of dairy bull semen.

The farm

Apart from keeping an eye on the sheep, I have done very little farming since I retired. Although there is plenty of time, it just isn't economic to invest and manage an enterprise of my own. It would also take too much physical effort. Besides, my current arrangement with Stuart Scott, the local farmer who has the grazing tenancy works very well. He does all the graft and I ride round on my quad bike and pretend I'm a working farmer.

As soon as I retired I was keen to do some improving of the eight acres of woodland and spent many happy hours with a chain saw tidying up the plantations and cutting out dead trees. When we first came to the farm in the early sixties, there was a four-acre wood in the middle of the land owned by the Water Board with access through the farm for their officials. In the early 1900s a series of natural springs in the area had been used to collect and pump fresh water to the village of Ashton, about two miles away. Long since replaced by mains supply, the concrete channels and small pump house remained within this four-acre wood, now hidden by overgrown vegetation. The Water Board decided to rid themselves of all of their small bits of such land and asked me if I was interested in buying this particular parcel. They accepted my modest offer of £700 and this opened up the possibility of felling some alder trees and creating a small lake in the middle of the site, fed by the spring water. In 1997 we cleared the area and employed a contractor to dig a hole about a quarter of an acre in size. Fortunately the ground was low lying and the water table obviated the need to line the bottom. Over the next few years we planted shrubs and water-loving species on the surrounds and even stocked it with rainbow trout. It is now a sublime oasis about a quarter of a mile from the farm and completely surrounded by a wide variety of trees. I visit most days, whatever the weather, and I hope it stays in the family for many years to come.

Farming at Spy Hill has always been extensive, although I did use fertiliser and sprays when we reared heifers and had beef cattle. I have since joined the Environmental Stewardship Scheme run by Natural England. This is designed to encourage farmers to conserve wildlife, enhance the landscape and improve water quality. Apart from some spot spraying to keep nettles and thistles under control the farm receives no fertiliser or herbicide. All the land is in permanent pasture and grazed by sheep. Hedges are carefully managed and only trimmed in alternate years to encourage nesting. When we first came to the farm we had dreadful problems with ragwort, a potentially poisonous and unsightly weed which flourishes on light land. We tried pulling and spraying without much success. Now with close grazing by sheep in the spring it has disappeared. This is a simple example of a management solution rather than a chemical one.

Over the last few years we have planted over 400 trees of different varieties in five locations on the farm and developed habitats around the lake to encourage new plant and animal species. We have also erected boxes for bird-nesting and bat-roosting. Barbara attracts an increasing variety of species through feeding and her generosity is well known by the bird population to be the best in the neighbourhood. Conservation and resource management are here to stay and we are enthusiastic to support any initiative provided it is not hidebound by red tape.

CHAPTER NINE

The New Millennium

THE APPROACH OF THE NEW MILLENNIUM had, for some time, been viewed as a potential computer nightmare as it was feared that many systems simply couldn't cope with the date change to 2000. Millions of pounds were spent exploring alternative arrangements to avoid the widespread chaos had there been a major problem. Like a lot of these scares, nothing untoward happened.

The New Year celebrations in London were not an auspicious start. The Royal party at the Dome was a flop; the Millennium Bridge over the Thames opened and closed quickly as it was considered a safety hazard and the much vaunted firework display to end all firework displays turned out to be a damp squib.

The first decade was memorable for turmoil rather than peaceful progress. The 9/11 attack by al Qaeda terrorists in 2001 shocked the world and led directly to a bloody and bitter war in Afghanistan. This was soon to be followed by the war in Iraq justified by the reported existence of weapons of mass destruction and the assertion that these could be activated within forty-eight hours. At home the London bombings in 2005 brought the terrorist threat even closer.

Politically the Labour Government under Blair and finally Brown presided throughout the period, towards the end of which came the banking crisis followed by a worldwide recession. The Royal Bank of Scotland recorded an annual loss of £24 billion, the biggest in British corporate history. Both it and Lloyds Bank had to be baled out by the taxpayer. Bankers soon became public enemy number one as they continued to pay huge bonuses to senior staff. Politicians were also soon heading downhill in the popularity stakes as the *Daily Telegraph* exposed a massive scam involving MPs who had been fiddling their expenses for years. Some of the worst offenders ended up in jail. It would be unfair to blame the Labour Government for all these events as the Opposition was in disarray for most of the time while they searched for someone who might revive their fortunes. David Cameron eventually became leader at the end of 1995, but it wasn't until the General Election in May 2010 that he became Prime Minister reliant on Liberal Party support in a hung parliament. For how long the Coalition Government will last as they implement a massive cost cutting strategy designed to stabilize the national debt, is anyone's guess.

The farming industry had to cope with another devastating foot and mouth epidemic in 2001. Farming practice had changed over the thirty-four years since the last major outbreak. Where immediate action to slaughter and dispose of infected animals contained the outbreak then to mainly Shropshire and Cheshire, this time protocols were under the direct control of the EU Veterinary Standing Committee in Brussels. Needless to say valuable time was lost due to this additional bureaucracy. To make matters worse, many of the small abattoirs throughout the country had closed down in the intervening period so that cattle and sheep had to be transported over long distances, thus increasing the speed of spread of this highly infectious disease. The epidemic started in Northumberland and quickly spread to Essex. It lasted for eleven months and affected over 2,000 farms, the majority of which (840) were in Cumbria. Some 10 million cattle and sheep were killed and the total cost was estimated to be in the region of £8 billion. The human cost was immeasurable but likely to be similar to that experienced in 1967.

Looking ahead to what the new millennium has in store – the challenges seem to be immense. Political upheaval in the Middle East, terrorism atrocities, unrelieved starvation in Africa, financial crisis within the EU are all in the headlines as I write and are deep seated and complex issues. Nearer to home we have to consider priorities and balance with regard to managing the environment and its effect on global warming. We need to think about new technology and consider how this might be used safely to solve some of the world's most serious problems. There will be others we don't even yet know about. This memoir is no place to pontificate about the future, but I offer some of my own thoughts with regard to a few of the issues which have an agricultural relevance.

The European Union

It is sobering to think that we joined what was originally the 'Common Market' almost forty years ago. At the time it seemed a sensible move to further trade interests with our European neighbours and also to strengthen defence policy during the Cold War. Whatever economic benefits may have accrued to other industries, I cannot be alone in thinking that the agricultural sector has been disadvantaged. Political influence through the National Farmers Union has now virtually disappeared as policy matters are decided in Brussels. For dairy farmers this has been particularly galling as the scope for expansion to fill the UK deficit in dairy products was removed by the introduction of quotas in 1982. Since then the number of milk producers has dwindled to about 10,000, a fall of 90,000 since I first joined the MMB in 1962.

I have always had doubts about the wider concept of a European Union and as the political process has progressed, it seems to me that there has been precious little measurable benefit, a massive net cost to the country and a great deal of unnecessary legislation. Furthermore Britain's sovereignty has been steadily eroded and we have no opportunity to alter things. Politicians of all the major parties know that a referendum on the subject would be likely to support withdrawal and therefore continue to avoid any commitment to allow voters to express their views.

There are countless examples of corruption, incompetence and mismanagement within the whole bureaucratic nightmare and now the entire edifice seems to be about to topple over as an ever-increasing number of the weaker economies have to be propped up by the stronger ones. Monetary union won't work where there is such economic disparity between member states and it must surely be a matter of time before some countries have to revert to their own currencies. However strongly people feel about the EU it may not now be possible to leave legally and of course the effect on UK trade within Europe will always be cited as the reason we have no alternative but to stay in. Anyway the protagonists will argue – what is the alternative?

I have no ready answer but was interested to read about a different slant in an article recently by Peter Oborne of the *Daily Telegraph* advocating a revamping of the Commonwealth as a means to counter the inevitable emergence of China as the global superpower of the future. Another potential benefit might be to create a credible alternative to over-reliance on our special relationship with the US. He suggested a relocation of the headquarters from London perhaps to India, nearer to the rapidly developing new world economies. Such a move would also help to distance the colonial history of its origin. The Commonwealth includes fifty-three nation states; 50% of its 1.7 billion people are under twenty-five, and apparently it costs a tiny fraction of that currently incurred by the EU to administrate. It is not tied down with legal constraints and it is an ideal vehicle to promote democracy in a quiet pragmatic way. If Britain truly wants to influence future global politics, it would do well to consider such an alternative rather than persevere with a half-hearted membership of the EU.

The environment

There have been dramatic changes for the good over the past fifty years with regard to the environment. The first two priorities were to improve the quality of air and water. I can well remember the appalling winter smog (a toxic mixture of smoke and fog) in cities and large towns when open coal fires were the main source of heating during cold weather. Smog led directly to breathing problems for hundreds of thousands of people. I referred earlier to rowing on the Tyne, where the quality of the water was little better than that of an open sewer. Now it is one of the major salmon rivers in the north-east. The sea was the dumping ground for all manner of waste and many beaches were unsafe for bathing – now they are closely monitored and much improved. There can be no doubt that such advances have been of tremendous benefit to everyone. More recently the subject has become a major political issue throughout the world. People are now more aware of the dangers and the complacency which allowed indiscriminate disposal of human and industrial waste without any regard to the consequences for future generations. Farming came in for significant criticism over the same period as a contributor to pollution, unacceptable animal welfare and practices which threatened wildlife habitats. With the benefit of hindsight, much of it was justified.

When I started work as a farm management consultant in the early sixties, there was little thought given to the potential side effects of intensive farming. Government policy encouraged farmers to expand production to save imports of food. The mantra of the day was 'Food From Our Own Resources'. There were grants, generous tax allowances on capital expenditure and subsidies on fertilisers to increase production. Dairy cows were intensively stocked and the key success criterion was profit per acre. The resulting problems in the dairy sector soon became apparent. On the most heavily stocked farms soil became poached, nitrates and slurry escaped into the watercourses, diseases such as mastitis increased in overcrowded buildings, and hedges were removed to facilitate the cutting of forage by bigger machines. No doubt I contributed to this trend in a small way as my first enterprise at Spy Hill Farm in 1965 was rearing dairy heifer replacements so that some of my farmer clients could increase their profitability by replacing their young stock with more cows.

Similar problems arose on arable farms with continuous cereal cropping, indiscriminate straw burning, and widespread use of herbicides. Non land-using enterprises such as rearing calves for veal, intensive pigs and battery hens all came under scrutiny as did livestock exports and the revelation that animals were being transported hundreds of miles without adequate food or water. Antibiotics and growth hormones were also being used indiscriminately to improve feed efficiency. The media focussed attention on the worst examples of bad practice and perception of the industry began to decline rapidly. It was also the start of green politics with the emergence of various organisations committed to galvanising public opinion and forcing the politicians both in Britain and Europe to introduce legislation to change the situation. At the extreme, certain groups began to take the law into their own hands against both individuals and organisations who were involved in agricultural development and research. Even Genus had a number of AI buildings and laboratories ransacked and burnt down.

One of the drivers towards intensification had been the system of financial support available to farmers throughout the EU. In 2003 following reform of the Common Agricultural Policy, a scheme based on eligible land area rather than production was introduced. In order to qualify, farmers had to conform to standards designed to protect the environment as well as plant and animal health. If they failed to meet these standards their payments were reduced or removed – a major incentive to conform. There is no doubt that this has been a major step forward compared to previous support schemes which paid producers for unwanted surpluses and even for not growing anything at all. The downside now is the endless legislative measures churning out from Brussels designed to control virtually every activity on the farm. Many countries ignore most of these, but the UK seems to follow them to the letter and employs armies of administrators to carry out farm inspections. The cost and the bureaucracy involved are exponential.

So the farming industry with regard to the environment has and is continuing to change for the better. Standards are policed rigorously, more so than on the continent. But I suspect the general public still feel that there is much yet to do and food scares are the result of continued irresponsible practice. Unfortunately food scares will always occur and the media will continue to be quick to locate the extreme view to fuel the story. I find it disappointing that they too often seem to exaggerate the negative side of farming rather than an attempt to provide a more balanced view. Maybe I'm over sensitive and I shouldn't be surprised as 'balanced views' don't sell newspapers The remark that 'you never see a poor farmer' or 'they all drive 4x4s' continues to be rolled out whenever I become involved in discussions with non-farming friends. Maybe this is tongue in cheek for my benefit and when pressed, most will agree that farmers haven't fallen as far yet in the public estimation as bankers or politicians. But I fear Joe Public still thinks they are feather bedded, they should have less financial support and that they should clean up their act.

Organic farming has become a popular option for those who feel that all chemical and biological additives should be eliminated from food production. The subject remains an emotive one with many of the protagonists claiming that the resulting products are safer and of higher quality. Not everyone agrees, but what is certain is the unit cost of production is significantly higher, partly due to lower productivity, but also to packaging and labelling. These costs have to be passed on to the consumer. There is a niche market but I think it will struggle to expand during times when the economy is under pressure. People have become accustomed to cheap food and despite media scare stories, are unlikely to pay the additional price required to make organic production a viable alternative to more conventional farming.

Of course there are bad farmers, as there are bad policemen or teachers and they should be weeded out. But, in my view, the vast majority want to produce high quality, competitively priced food, look after the environment and make a reasonable profit. Unless they manage their land and livestock properly and responsibly they know they won't achieve any of these objectives and that public perception will fail to improve.

New technology

Genus is now at the forefront of new reproductive technology within the cattle and pig sectors of animal breeding. Selection has been practised for years in an attempt to improve the capacity of domestic animals to produce more milk or meat for the same input of food. Until the arrival of AI in the 1940s, progress was slow and uncertain – more art than science. It was little more than natural selection based on observation of the parent's performance. Artificial insemination opened up massive opportunities, although at the time the procedure was deemed unnatural and there was significant opposition by church leaders in particular. Now, such a technique is commonplace, but the science moves ever forward searching for new ways to shorten the length of time it takes to achieve genetic improvement. Embryo transfer, semen sexing and pregnancy scanning are all now techniques in widespread use with the aim of producing more from less.

The wider and more controversial topic of genetic engineering is another which attracts much debate. Genetically modified organisms (GMOs) are organisms such as plants, animals and micro-organisms (bacteria and viruses, etc.), the genetic characteristics of which have been modified artificially in order to give them a new property – for example in plants – resistance to disease, insect infestation, cold and drought. The potential to increase food production in many developing countries is immense. In animals, genetic engineering can be used to eliminate some congenital diseases and to increase productivity. But there are understandable and justifiable concerns. What are the long-term effects on plant, animal and human health? Is it possible that accidental cross-breeding will occur so that non-targeted species such as weeds also inherit the same benefits of resistance? The antagonists of such technologies argue that no one knows the answers and the extremists advocate that all trial work should be banned. It is a contentious and highly charged debate and one which will stretch well into the future. Strict regulation is essential and rigorous testing of new products must be a prerequisite of their release onto the commercial market. But in my view, subsequent generations would be rightly critical if scientific advances in this field were left unexploited. They have a potential to solve many of the world's intractable problems.

Global warming

The science behind the hype depends on the observation that there continues to be an increase in the average temperature of the earth's atmosphere and oceans. It is caused by increased concentrations of greenhouse gases (water vapour, carbon dioxide, methane and nitrous oxide) resulting from human activities such as deforestation and the burning of fossil fuels. This conclusion is recognised by the national science academies of all the major industrial countries and is apparently undisputed by any scientific body of national or international standing. Temperature records show that average global surface temperature increased by 0.74 degrees centigrade during the twentieth century. Annual variation over this period has been pronounced and more recently in the first decade of the new millennium there has actually been a levelling trend. The Intergovernmental Panel on Climate Change (IPCC) forecasts a further 1.8 to 4 degree increase during the twenty-first century. This is where the science ends and the speculation begins. Computer modelling is the methodology involved and it is easy to see where the scare stories start: with massive ice melt, rising sea levels and extreme weather patterns based on the higher numbers in the range. No one knows what will happen but in any event it makes sense to try and reduce or at least stabilise greenhouse gas emissions.

The subject is high on most political agendas throughout the world as it should be. The aim is to achieve consensus and commitment for individual countries to take action to reduce gas emissions as well as undertake programmes of geo-engineering in an attempt to reflect incoming solar radiation back into space. The costs of such programmes will be enormous and the new world economies are unlikely to agree that the pain ought to be distributed equally. They will reasonably argue that industrialisation in the West has been the main cause of the problem and their populations should be allowed leeway to catch up.

The most interesting book I have read based on the available evidence is by Nigel Lawson, a former Chancellor of the Exchequer and now ennobled as Lord Blaby. In *An Appeal to Reason – a cool look at global warning*, Lawson debunks much of the dodgy science and media hype. Of course there are actions that need to be taken and he summarises these in a practical common sense way. Antony Jay, the celebrated writer, is more forthright. He maintains that no one in authority should pronounce on global warming again until they have read this book. He opines:

> At last we have an independent hard-headed examination of the realities of global warming. Nigel Lawson slices through the layers of pseudo-scientific hype, anti-American prejudice, green evangelism and rampant ecomania to expose the scientific realities, the political issues, the economic options and the ethical considerations that really matter. He brings to the debate a rare breath of intellectual rigour, political experience and sheer common sense that have been missing for far too long.

Global warming is but the most recent international scare story. In the mid-seventies the planet's temperature started to fall and we were warned to prepare for another ice age, similar to the last one in Britain 400 years ago.

Population experts point to increases of 40% to over 9 billion people by 2050, mostly in developing countries. This is mainly due to advances in medical science and the steady increase in life expectancy coupled with reduced infant mortality. Birth rates may fall but the need for the world to produce more food is indisputable. This is not an impossible challenge, provided technology and mechanisation are put to good use.

Next came the warning that the world was fast running out of natural resources and that within our lifetime economic growth would end. The search for alternative and renewable energy sources using solar, wind or tidal power is gathering momentum as is the debate on the expansion of nuclear power in the light of recent disasters such as Chernobyl and the recent Japanese tsunami. One thing is certain, continued human adaptability to global warming, as in everything else, will happen. There will be new discoveries and inventions within the competitive economy which will occur without the need for government action.

Towards the end of his book, Lawson highlights the dangers that the new religion of eco-fundamentalism and global warming presents:

The first is that it breeds an intolerance of dissent and reasoned argument that is both unattractive and dangerous. The attempt by the Royal Society, of all bodies, to prevent the funding of groups and organisations which openly doubt the alarmist creed of the new orthodoxy, on the grounds that they are providing inaccurate and misleading information to the public is particularly shocking and telling. It is clearly undesirable that no young scientist, or young politician, dare question the new religion without severely damaging their career prospects. It is no coincidence that those who do are mostly retired.

The second danger is that the Governments of Europe may get so carried away by their own rhetoric as to impose measures which do serious harm to their economies. This is a particular danger at the present time in the UK.

And the third, and still greater danger, is that even if the voters prevent Europe's governments from going too far to damage their own economies, they may still cause great damage to the developing world by engaging in what might be termed green protectionism. The movement to make us feel guilty about buying overseas produce because of the 'food miles' involved is just one example of this.

Public opinion within a democracy should result in political decisions which the majority support. There are millions of people who either don't care or can't be bothered to inform themselves of these and many other issues which affect their lives and the lives of their children now and in the future. Those of us who have been lucky enough to have had a good education have a responsibility to think, read and talk about these problems and the possible solutions, rather than just believe what we observe and see in the media. This is the only way to formulate a valued judgement and as a result influence other people's opinion. It is not an option – it is a duty.

CHAPTER TEN

Retirement – and Second Childhood

SOME PEOPLE WOULD HAVE YOU BELIEVE that your schooldays are the happiest of your life. Being a student came close but the last decade would be my preference. Between sixty and seventy, there exists an exciting window of opportunity for those who have the means and health to exploit it. Thereafter the body begins to object to the pace of life and you have to think and plan more carefully as inevitably you approach what Shakespeare defined in his seven ages of man as 'second childishness, and mere oblivion,/Sans teeth, sans eyes, sans taste, sans everything.'

I have always been keen to have a project on the go, something to plan and think about. Creating the lake on the farm fell into this category and involved a good deal of physical effort which I enjoyed immensely. The next one was to be very different – the writing of a book, a biography of Mac Cooper, Barbara's father.

Mac had been encouraged by his wife, Hilary, to write an autobiography and he started enthusiastically. However, he didn't get further than his student days as he disliked using the personal pronoun. Fortunately his original draft survived and described his childhood on the family farm in New Zealand and his student days at Massey Agricultural College. The manuscript was given to me by the family to consider after he died in 1989. It remained on the backburner until I retired seven years later, by which time Hilary had also died and with her, a valuable source of information. However, I now had the time, the inclination and the encouragement of their three daughters to get on with it. I had written lots of articles and papers during my career, but nothing as remotely ambitious as a book.

I referred briefly in Chapter 2 to Mac Cooper's background and reputation when he arrived at Newcastle as Dean of Agriculture and where, in 1956, I became one of his students. Later as one of his sons-in-law, I got to know him well as a person, but it was clear that if I was to do a proper job, then I had to undertake some serious research.

I started in earnest in the spring of 1999 by visiting New Zealand for the specific purpose of meeting family members, former colleagues and friends of Mac's in order to try and put together a picture of this exceptional man. One of the main discoveries was to be a comprehensive collection of letters from Mac to his sister, Thelma, who, nearing her centenary, was the only surviving member of Mac's six siblings. He had been the youngest in the family and had always had a special relationship with the younger of his two sisters. This correspondence was a unique source of information, beautifully written in his own inimitable style, often emotional, always humorous and honest. These letters were Mac's means of keeping the family informed of his activities when he was overseas and they provided an insight into parts of his life that no one else, apart from Hilary, knew about. They spanned the duration of his war service in Italy and his career as it developed in the UK from 1947 when, at the age of thirty-seven, he accepted the post of Professor of Agriculture at Wye College in Kent. Apart from occasional visits he continued to live in the UK for the rest of his life. Thelma was thrilled that the project was underway and for obvious reasons, anxious that I should not delay. She was happy to lend me her collection and agreed that I could quote from these letters extensively when I came to write the narrative.

Mac was born in 1910 and grew up on the family farm near Havelock North in the North Island of New Zealand. He went to Massey Agricultural College as a student and stayed on briefly as a lecturer before the war. I visited his school, his various homes and Massey on my research trip and met people who knew him well at that time. I also visited other New Zealanders, some of whom had been with him in Spain where he had worked briefly for the World Bank as National Research Co-ordinator after leaving Newcastle in 1971. I met more distant relations throughout the North Island and left, overflowing with material, after a very intensive twelve days. Back in the UK I went to Wye College where the Principal, John Prescott, had been one of Mac's students at Newcastle. He arranged access to the College archives and I was also able to talk to some of the older staff who remembered Mac well. I visited the Rhodes Scholarship Trust in Oxford where I was allowed access to Mac's personal file which contained some priceless exchanges of correspondence between the academic staff about the anticipated arrival of this 'new Rhodes Scholar from New Z. who has taken a degree in Agricultural Science and whose speciality is pig feeding!' I spent a lot of time at Newcastle where memories were more recent and also met some of his former colleagues, now retired. Not all of them were members of his fan club as he had arrived like a bolt from the blue in 1954 to re-galvanize the faculty. There had been inevitable casualties. I hadn't seen many of these people for over forty years when, as a rather noisy and rebellious student, they had given lectures. They were very polite and helpful but couldn't disguise their amazement that I had eventually graduated let alone embarked on the task of writing a book about their former boss.

The book was eventually published at the end of 2000 and the first twenty copies arrived when Barbara and I were on holiday in New Zealand, enabling us to give copies to Mac's family during the course of our visit. I was particularly anxious to present Thelma with hers when we there. It was a close run thing. I had made arrangements with the publisher to have the first copies available to take with us but they were delayed, although thankfully arrived at our first port of call when we got there. I didn't know then that the publishers were about to go bust within a few months and my pleas to meet a deadline were probably low on their list of priorities. I had 500 copies printed and gave away about 100 of these to family and friends. About 350 were actually sold before the publishers went out of business, although I didn't receive all the money. It wasn't a commercial success but then that wasn't the objective. The family were delighted with the result and I received many letters expressing both surprise and satisfaction that the book was a fair and affectionate portrait of an inspiring man, warts and all.

I enjoyed this project and am happy in the knowledge that Mac's life and legacy have been recorded for posterity. Amazingly the book still appears on the Amazon search engine under 'Mac Cooper' with a few used copies for sale at more than the retail price. It must now rank as a collector's item.

A Ride to Remember

The next project took place in August 2001. A very good rugby friend of mine, Ted Charlesworth, had been diagnosed with Parkinson's disease and he encouraged a group of us to ride a bike from Land's End to John O' Groats. I happily took on the role of organiser and chief cashier.

Ted had been head of the PE Department at Chester College, having qualified as a teacher from Loughborough in the Fifties. He was a gifted athlete and played both cricket and rugby for Loughborough as well as representing Yorkshire at rugby. He was just finishing his first team rugby career with the Chester club when I joined and we played a couple of times together before he retired to become the club's first coach. We then played golf together for many years at Chester Golf Club.

Ted was a fitness fanatic and probably one of the most enthusiastic competitors I have ever come across. It was, therefore, a tremendous shock when, in 1992, he was told he had the early onset of Parkinson's. This was only the start. He had hoped a couple of years later to have a knee replacement before he learned that a heart bypass was a more pressing priority. Eventually he died from cancer in 2002 at the age of seventy-two. For many people a catalogue of health problems such as these would have reduced them to despair and depression. Not so with Ted. I quote from a recorded conversation I had with him before writing an account of Project Ted following the bike ride:

> I do have low periods but perhaps not so many as I ought to have. At the back of my mind I feel it might get better so I don't think of myself as a hopeless prospect. The pills work and they will produce new pills, so you aren't thinking this is the end – you feel there is some hope. In the early stages I was trying to hide things. I didn't want people to know that I had some terrible illness. But you get to the stage where you don't mind. I almost feel quite proud sometimes that I can do things people think I can't do. So I don't get depressed – I get low occasionally like anyone else gets low, but so far I haven't had what you would call depression.

This gives some insight into Ted's strength of character. He had a zest for life, irrespective of the hurdles that seemed to crop up along the way. There is no element of self-pity or 'Why should this happen to me?' How often do we see people with a fraction of his burden take that view, making life so much more bearable for their carers, friends and loved ones. He was an inspirational man.

The background to the project was the revelation that Ted discovered, almost by accident, that he could still ride a bike. Some of his friends decided to buy him one for his seventieth birthday and as a result a group of us used to meet on Tuesdays and cycle round the Cheshire lanes, stopping for lunch at a local pub. In spite of his decreasing mobility, he was amazed to find that pedalling and balance were unaffected. He recalled: 'the joy in being able to sweep along at the same speed as everyone else, uncluttered with these legs that won't move and get all knotted up, is unbounded.' One day when we had been rather longer in the pub than on the bike, the conversation turned to whether we might embark on a rather more ambitious expedition, perhaps to North Wales. Ted proposed something 'memorable while he still had time – how about Land's End to John O' Groats (known colloquially as Le Jog) and maybe raising some sponsorship along the way?'

In the sober light of the next day we decided to meet and see if we could produce a plan which was both feasible and achievable for a bunch of oldies to tackle this formidable challenge. Fortunately a mutual golf club friend, David Faulkner, was the MD of the local *Chester Chronicle* and thought that the concept of a group of sixty-somethings riding from one end of the country to the other in aid of the Parkinson's Society with Ted as the focal point had real market potential. He promised to produce all the publicity material for us and we were committed. We even had Michael Owen, the footballer, to do a photo-shoot with some of the team at the golf club, where he was also a member.

We decided to ride the bike in relay for the 874 miles from end to end, making the wise decision to start from Land's End as the prevailing wind was from the south-west. We would take it in turns to cover about eighty miles per day. The broad format was to have four riders on each day, one of whom would be in the saddle and one in the back-up car during the morning session with the other two having time off. We would then change round in the afternoon. This way each rider would pedal about twenty miles per day. We would stay at pre-arranged B&Bs and fund all the costs ourselves. We would also aim to raise as much as possible in aid of the Parkinson Society, with whom we had agreed that the money would contribute to the specific aim of funding a specialist nurse for the Chester area.

As the concept moved from theory into practice, we had to recruit a team to undertake the task. The members of the Tuesday Club by this time had drifted into two groups – the A Team who started to up the tempo, increase mileages, buy Lycra shorts and generally hone their bodies for the forthcoming challenge. There were four of us, including myself, who fell easily into this category. Brian McAdam, another former rugby player at Chester was also into golf and had unearthed his ancient bike to join our activities on Tuesdays. A born salesman – he had worked in the chemical industry and retired at sixty-two. He was an arch enthusiast, full of energy, loved the ladies, was somewhat forgetful and prone at times to lose his cool. A great character and exciting company. Tom Dewhurst at seventy was older than most of us but just as game. A Cambridge graduate and football blue, he was also a keen cricketer. A former teacher, he had recently been the grateful recipient of two new knees so that he could now stride around the golf course and pedal a bicycle with the energy of a man twenty years his junior. Rather more cautious than some of his fellow team members, Tom realised the enormity of what we were letting ourselves in for and counselled realism about some of our more ambitious targets. Mike Foster, the youngster of the group at only fifty-seven, had recently taken early retirement from Unilever having been a senior Market Research Executive, specialising in environmental issues. Mike was probably the most capable golfer in the team as well as preferring the round to the oval ball. Somewhat more reserved than some members, he was a mainstay of reliability and determination.

But four serious riders were insufficient, we needed two more, which I was detailed to find. One obvious candidate was former second row forward Joe Bright. He had recently retired from heading up a Danish firm which made plastic bottles in Runcorn. At sixty-plus and pretty fit, Joe was just the man. I broached the subject midway through a convivial day as his guest at Chester Races and he was hooked without bother. He was a proud Liverpudlian and one of the funniest men around. His involvement guaranteed some hectic nights in the furthest flung corners of this sceptred isle. The final A Team man was Mike Pott. Although neither a member of Chester Rugby or Golf Club, I knew him pretty well as he was Pete's father-in-law. He had recently retired from running a pharmacy business on the outskirts of Chester and enjoyed outdoor pursuits such as walking, cycling and golf. A fit mid-sixties man, he agreed to juggle his diary enabling him to join us for the final six days of the trip in Scotland.

The B Team was more content to limit its Tuesday routes to the flat plains of the county and discover new hostelries. It was clear who was going to have to cover the hard miles. We managed to persuade six people to join the project and contribute relatively small mileages within a reasonable distance from Chester. Jeffery Huckerby was the leading light in the B Team and although he would not divulge his age, we knew he added significantly to the average. A close friend of Ted's, Jeffery is another of life's natural extroverts, always at the centre of activities and a great asset to any team. Ken Hayton, a retired Chief Fire Officer, would have loved to have joined the A Team but he was having a few health tests and his wife, Jackie, forbade him to ride. He acted as our quartermaster and arranged all our accommodation throughout the trip. He also managed to jump on the bike when Jackie's back was turned. Four other sporting friends completed the B Team: Brian Hayes, one of the biggest scrum halves ever to play for the 1st XV; Mike Hattersley, a substantial prop forward, Robin Bates, an elusive centre three-quarter and Chris Cox whose exploits with a hockey stick were renowned throughout the north-west of England. We had sufficient manpower, although much to his disappointment, Ted was unable to handle a bike by the time we came to start the ride. He came along in the support car for most of the time and joined the Scottish leg with Chris his wife, and old friends Ken and Jackie Hayton.

We rode the bike for 1,024 miles, some 150 further than the published route. This was partly due to seeking quieter and less dangerous roads, diverting to take advantage of friends who offered us free accommodation and sheer incompetence in map reading. The A Team between them covered 91% of the distance, the B Team did the rest, although if you talked to them they might persuade you they did much more. The trip took thirteen days, five from Land's End to Chester where we had the weekend off and eight days to JO' G. We averaged 10.4 m.p.h. and the record speed achieved was a hairy 44 m.p.h. – down a very steep hill in the north of Scotland dodging the sheep. In total we raised just over £30,000 for the Parkinson Society.

It was a marvellous experience and all of us will remember it for as long as we live. I kept a detailed diary of our exploits within which most of the team added comments as we wended our way northwards. I later edited this diary into a booklet which we sold at a price of £2 to add to our funds. There is not room in this book to record all our experiences, apart maybe one episode which occurred shortly after we had crossed the border into Scotland.

> The welcome awaiting Jeffery as he passed the sign for Scotland would have gladdened Sir Walter Scott's heart. The fan club had expanded to seven all waving and cheering on their man as he pedalled gamely up the final stretch. Another change and Mike Hattersley climbed aboard the bike to power his way towards Langholm and lunch. Now here, as in many recollections, there was a range of views as to (as they say) what happened next.

Mike records in the diary:

> After about 3 miles the outlook changed. The road narrowed, the downhill sections were short (I was told the whole length was downhill),
> the uphill sections got steeper and longer, so by the time I got to Langholm (all of 8 miles) I was feeling somewhat cream-crackered. Having overcome a slight difficulty dismounting the bike we then went into the town for lunch.

Jeffery who was in support and gave Mike his directions, reports the arrival of the 'Queen of Sheba' in more extravagant terms:

> Our meeting place was a large picnic area on the banks of the river. Mike was greeted to sympathetic cheers from everyone, including a small child in a pram who had become quite frightened by the sight of grown men in tears. Mike was lifted off his cycle with difficulty and only came round somewhat abruptly after a totally unnecessary 'kiss of life' from the Chief Executive.

I wrote:

> It is probably true to say that Mike's leg was rather more testing than most of

us realised. In fact his problem was his leg. He couldn't get it over (backwards that is) the saddle so that he could dismount with dignity. Four attempts failed miserably and like a drowning man he sank, bike and all onto the grass, to be helped to his feet by his hysterical team mates and their attendant fan club.

When we were within striking distance of the summit, I was given the honour of cycling the last few miles. I recorded the following in the diary:

> It was with mixed emotions that in bright sunshine I freewheeled the last couple of miles through the village of John O' Groats and down to the jetty from which the ferry left regularly for the Orkney Islands which could be clearly seen in the middle distance. A little sad that our epic effort was at an end; relief that it had been accomplished without accident, illness or injury – no mean feat considering our ages and a sense of achievement in that all the planning, fund-raising and execution of the ride had been manifestly successful. And finally, immense satisfaction that Ted himself, had derived so much enjoyment from the project.

Spy Hill Lodge

The most recent project (apart from this memoir) and the most important has been the planning, building and moving into our new home across the yard. Barbara and I had been thinking about the longer-term future for some time, realising that sooner or later we would have to decide whether to continue living at the farm or sell up and find somewhere smaller and easier to manage.

The thought of leaving the farm was a subject too easily put to the back of the mind. But we also agreed that when one day the survivor, whoever it might be, would not wish to continue living alone in the house. A possible solution began to emerge when, in discussion with Peter, he mentioned that they were considering a move to somewhere with more space. In fact he had already looked at some places locally. Was it possible to put together a plan for him to buy the house and for us to convert one of the farm buildings into a one-storey, modern dwelling? We discussed the idea with the other members of the family and they all agreed this seemed a sensible option and so we set off on the long and tortuous journey to obtain planning permission to convert the old cowshed.

I now understand why so many people despair and give up the unequal fight with the local planning bureaucrats who seem to make life as difficult as possible. According to the consultant we hired to help us prepare our case, these people are creating more and more hoops for the hapless applicant to jump through. It seems in many cases to simply help preserve the jobs of various hangers on. One example is the requirement for bat surveys in rural locations which require specialist consultants, recording data, lengthy reports, large fees and EU licences, if the evidence points to the presence of a few bats roosting in the buildings. Bat boxes and even specialist facilities on site have to be erected to house these creatures before work can start. The discovery of newts requires even more draconian measures to avoid disturbing their habitat. It seems crazy in this day and age that this sort of restriction is allowed to delay and sometimes prevent housing developments to proceed while small builders, desperate for work, continue to go bust. I am well aware of the need for responsible development and animal welfare but this sort of regulation assumes that animals are incapable of finding new habitats. They have done so since time began.

Anyway we persevered and eventually got the necessary permission to proceed and employed a small family builder from Wrexham to do the job. They and their subcontractors commenced in the autumn of 2009 and did a splendid job, enabling us to move in during August the following year.

The original line of the old building and lean-to remains, the roof having been replaced and under-floor central heating installed. All the living accommodation is on one level but with extensive loft storage above. With two bedrooms, open-plan kitchen and garden room together with a large lounge and office, the layout fully meets our needs. Outside we have landscaped the immediate surrounds and installed a large patio area with grass and shrubbery features. It is a tranquil place and gives us peace of mind about the future, whatever that might bring.

Other pursuits

Rowing and rugby featured prominently as my main sporting interests as a young man. As the body gets older, one has to adapt to less strenuous activities or simply just become a spectator – not something I wanted to do. I had tried tennis and squash without much success. We took up skiing in our mid-forties and whilst not very proficient, enjoyed many uplifting holidays in the Alps during the winter.

I had a few goes at sailing with various friends. One trip on board John Willis's 38 ft. boat *Rapid Transit* was an unforgettable experience, especially as the other two crewmen were Joe Bright and Brian McAdam of 'Le Jog' fame. This was a fortnight's voyage in July 2002 starting from Caernarvon in North Wales and visiting the Isles of Scilly, Brittany and the Channel Islands. John, a golfing friend and former rugby player of considerable proportions, was an experienced sailor and Joe was also well versed in matters of the sea. Brian and I were virtually complete beginners. The first leg, overnight, was quite rough and most of the Irish stew we had enjoyed earlier in the evening ended up in the sea of the same name. I began to wonder whether this was such a good idea. We kept a daily log of our progress and my first entry records some of the hazards:

> There were always two of us on watch during the night lasting 3 hour stretches. Joe and I had the 8 to 11pm and the 2 to 5am slots last night and they were a new experience for me. It was a distinct advantage to be up on deck rather than down below as the pitching and rolling meant that you had to hang on for dear life if you were to avoid being thrown about. The trick was to get into your sleeping bag as quickly as possible once you were off watch and lie down. Amazingly once horizontal I felt OK. We shared a cabin amidships – me on the top bunk and him below. The 'en suite' facilities were just across the corridor but a major feat to negotiate in rough weather. The 'head' is a primitive sit down loo, the contents of which, after use, have to be pumped into the bilge. In practice you needed three hands. One to hold up the seat, one to hang onto the roof as each wave sent you rocketing side-ways and one to try and guide most of it into the bowl rather than on to the floor. It might have been easier if we had been female, but I'm not too sure. I clearly have a weaker bladder than Joe and despite limited liquid intake, needed to clamber down from the top bunk during our brief period of sleep. This meant he had the pleasure of my foot landing somewhere on or about his person as I made the descent and when I returned for the ascent – but he never complained – a real sailor.

For most of the trip we had glorious weather but the wind always seemed to be against us which meant using the engine more than intended. For me, the dawn gradually coming up over the horizon, the sea birds relentlessly in our wake, the smell of the sea and the realisation of your insignificance in such a vast expanse of water were the highlights. I can understand why some people are constantly drawn back to the sea but I'm not one of them.

Golf seemed to be the preferred option for many of my rugby friends and whilst I had played with my father on and off from the age of about ten, this seemed to be something worth taking a bit more seriously, despite Mark Twain describing it as 'spoiling a good walk'. It's the only game I know where a good golfer can have a properly competitive game with a poor player – through the handicap system. You can also play it well into old age, as some of my current golf partners can ably demonstrate.

I joined Chester Golf Club in 1980 and have enjoyed over thirty years of membership, making new friends and whiling away the hours swinging a club. Averaging over two games per week I reckon I must have walked the equivalent of well over the distance between here and New Zealand in pursuit of a little white ball. In those days you couldn't just walk in and join a club. Many had long waiting lists and interviews with the captain and secretary to check out whether you were the right sort of chap. Maybe my rugby connections helped and I was soon part of a group who came to be known as the 'Teddy Bears', appropriately named after their organiser and guru, Ted Charlesworth.

We would meet every Wednesday and Saturday morning, draw for partners and then play a round, being credited with points depending on the result. At the end of the season the declared winner won the teddy bear, although no one ever knew how Ted calculated the results and it didn't matter very much anyway. I even won it one year having played only a handful of times. Such an activity was not altogether approved of by some of the senior members of the club at that time as it was seen to be a club within a club. It was better to mix and meet new people. Nevertheless currently there must be ten different 'roll up groups' who do broadly the same thing.

Now we tend to play twice a week in a regular four ball. One of the group is Jeffery Huckerby who captained the B team during the cycle ride in aid of Parkinson's Society. He is about ten years older than me and our golf could be described as usually modest and occasionally exceptional. We are both very competitive and hate having to award the £1 prize money to each other. Moreover neither of us is averse to a little gamesmanship to gain advantage. Every time I address the ball on our short 17th hole at Chester I can hear him behind me whispering to his partner 'I hope he avoids the ditch this time'. Invariably I don't which inspires a prolonged fit of giggling and my stalking off down the fairway vowing never to play with him again. We recently played at a rather up market and hilly course in Cheshire as members of a visiting society. We decided to see if we could hire a motorised buggy to ease the effort after having previously played a few games on consecutive days. We were told, rather pompously, there was only one available and we would need to provide medical evidence to justify our use of the four-wheeled, two-seat, petrol-driven, evil-smelling and noisy vehicle.

I forged a letter in my wife's name as follows:

TO WHOM IT MAY CONCERN

Dear Secretary

I am a retired General Practitioner and herewith provide my opinion as to the reasons why the two gentlemen named below require, largely for medical reasons, the use of a buggy at your golf course on Friday 26th August 2011.

JOHN CRAVEN
He is 74 years old and experiences the following difficulties:

1. Irregular heartbeat. His medication sometimes has unfortunate side effects which require his rapid transit to the men's changing room.
2. Partial sight in the right eye. There are frequent occasions when he loses sight of the ball in flight and the need for a second opinion as to its likely whereabouts.
3. Hearing inadequacies. He is unable to hear warning shouts from other golfers whose wayward shots arrive from different directions. The protection which the buggy provides may increase his safety and reduce your insurance liability.
4. Multiple teeth replacements due to grinding in hazards of which your course has many.
5. Chronic damage to the left ankle, causing sudden and dramatic falls, especially on uneven ground.
6. Mental fatigue and intemperate language when he plays a bad golf shot.

JEFFREY HUCKERBY
He is 85 years old next Thursday. He is remarkably well preserved, due in no small measure (a large one actually) to his addiction for an unusual cocktail based on gin and Chateau Neuf du Pape.
It is wise for him to use motorised transport whenever he plays golf, if for no other reason than he needs to sit down for a few minutes before he takes each shot.

Yours sincerely
Dr Barbara Craven

We have a lot of fun and I hope our games with friends continue for many years to come.

For many people golf clubs have a reputation for being rather stuffy, old fashioned and unwelcoming to visitors. This was certainly true of Chester in the Eighties and I well remember one summer's day being taken to task by a pompous little committee man for not, in his view, wearing 'tailored' shorts.

But times have changed in golf clubs as in everything else. There are twice as many clubs as there were and the waiting lists have disappeared as have joining fees – usually one year's subscription in advance. Virtually all clubs are competing against each other for members. Younger people now often prefer to play in societies on different courses when they want and at a cost which is lower than committing to the annual subscription that clubs now have to charge to remain viable. Running a golf club is today similar to managing a commercial business – it has to provide what the market wants. Some of the older members still rue the changes and yearn for the old days when standards were maintained and the class of membership was more selective. But like old soldiers they will never die – but simply fade away. It was ever thus.

Barbara also took up the game in her fifties. Having a wife who plays is of great benefit in golf as the game does take about four hours excluding any time which may be spent in the bar afterwards. There is, therefore, no need to have to continually explain this fact to spouses left at home and no guilt complex at leaving them on their own while you enjoy yourself. We both have our separate group of friends with whom we play regularly. Furthermore there is the obvious bonus of occasionally playing together, often on holiday with friends and also in mixed competitions. We have both been delighted to serve as captains of our respective sections and also play a part in the committee work which is an essential component of amateur golf clubs. Neither of us could be described as good golfers but we enjoy the exercise, the friendships and the social life which used to be centred on the rugby club but seems now to have moved on to the golf club.

Shortly after I retired I had a game with Duncan Spring and Andrew Dare, both people I had worked with in the past during my career with the MMB. They were also at the point of retirement and we decided to form a small golf society – for the want of a better name we called it the XVI Cub. The idea was to bring together a group of about sixteen people of similar vintage who we knew played golf and had also been prominent in their various fields within the dairy industry. We ended up with a mixture of ex-colleagues and dairy farmers from all over the UK and Ireland. It became an excellent opportunity to keep in touch with people you once worked with and who you may never have seen again had it not been for this idea. Our inaugural meeting was in Cheshire and it became so successful we expanded the numbers and invited the ladies to join us. We take it in turns to organise events and over the last fourteen years we have played at about forty different locations throughout the UK as well as in Ireland and France. The two main annual competitions are for the 'Wits and Shanker's Trophy, and the 'Double D Cup.' Barbara is the current holder of the latter, but sadly I haven't yet managed to get my hands on the former.

Whilst sporting activities have played a major role in my life, other interests and hobbies have also been important and the time available to enjoy them is another bonus of retirement. Reference to musical talent in the family has been made earlier, but unfortunately I didn't progress further than a few scales on the piano. Apart from a passing acquaintance between Pete and a trombone, the musical gene seems now to have skipped three generations, but we live in hope. I have always enjoyed music but for me it has to have a recognisable tune, not too loud and to include some harmony. Although not religious, I enjoy hymns, choral music and brass bands, as well as opera and classical music. We are fortunate to live within an hour of visiting either of two nationally renowned concert halls – The Bridgewater Hall in Manchester and the Liverpool Philharmonic. I admit to intolerance when at a function, the music becomes so loud that it is impossible to hold a conversation. I also admit to impatience with regard to much of the so-called entertainment on television. Maybe this stems from way back when occasionally I used to accompany my father on his evening rounds of the cinemas which were then the family business. Whilst he was checking the night's takings in the office, I would go into the auditorium to watch the film for about fifteen minutes. Then I was whisked away to the next venue where the same process was repeated. So unless I am enthralled by a TV programme within a quarter of an hour, I prefer to do something else. Rugby, golf, news, documentaries and 'Dad's Army' are notable exceptions. So I now read a great deal. The facility to obtain books, usually second-hand at a fraction of the retail price, is one of the great advantages of the internet. My preference is biography, military history and the period between the Industrial Revolution and the end of World War One.

A hobby I have enjoyed for over sixty years has been photography. My first camera was a Box Brownie and I learned in the dark room to develop the negatives, enlarge and print the black and white pictures using all the various chemicals then required. I have managed to acquire the family albums, many going back to the 1920s. They provide a graphic history of those early days and contain many images of long forgotten family and friends. I have a substantial collection from when the children were small, many in transparency format when colour photography first appeared. Now, of course, the digital camera has revolutionised the industry and less skill is required to record high quality images, although many still seem to make a mess of it. Albums containing prints are a marvellous way to record an event or a holiday, although I suspect an increasing number are now simply stored on disc rather than ending up in an album.

CHAPTER ELEVEN

A Family Update and Some Reflections

LOOKING BACK TO THE EARLY DAYS when the children were small, I know I missed too much of their upbringing. As I moved ahead in my career, the need to be away from home increased and this meant the extra burden fell on Barbara – at times I think she felt that ours was a 'one parent' family. I marvel now at the amount of time our children spend with theirs. Maybe they identified the priority better than I did – you only get one chance. One thing I did do was to ensure we always had a proper family holiday. When they were small and money was tight we went to places like Arran and Aberfeldy in Perthshire, climbing hills, fishing and staying in small hotels – sometimes rather stressful at mealtimes. Some friends of friends owned a villa in Menorca where we had three marvellous seaside holidays. Later we went camping in France and even hired a longboat on the Shannon, where it never stopped raining. We often kept diaries of those holidays with each member contributing a record for the day. Together with hundreds of photographs they provide a priceless memory of those holidays as well as illustrating hilarious sketches and jottings. They will sustain us in our old age.

Rugby also featured prominently in our shared activities. After retiring from playing at the age of thirty-six due to a combination of age and a bad back, I took up coaching and organising the junior section, which in the seventies was just beginning to be taken seriously at club level. The boys were all mustard-keen and it has been a joy to see them and indeed their own children get so much out of the game and make so many friends through it. Rugby is not just a game it is a brotherhood.

By the end of the nineties we had an extended family of twenty-one. Sue married Neville Sorrentino in May 1988. They met whilst at Rydal, where Sue joined the sixth form and after 'A' levels went up to Sunderland to study pharmacy. Neville graduated from Liverpool with a sports science degree but decided to start a wholesale carpet and furniture business in Llandudno. They lived in North Wales for some years, where their children were born and then decided to move to the British Virgin Islands in the Caribbean. Neville had

41. Spy Hill Farm looking east towards Delamere Forrest, Autumn 2010

42. Spy Hill Farm looking west towards the farm and buildings, Winter 2010

43. The lake, early Spring 2004

44. Excavating the lake, 1996

45. Ben Craven supervising operations

46. Members of the A and B Teams
Standing from the left: Chris Cox, Joe Bright, Mike Pott, Mike Foster, Brian McAdam, John Craven.
Sitting : Ken Hayton, Tom Dewhurst, Jeffery Huckerby, Ted Charlesworth, Robin Bates

47. A Ride to Remember. The start
at Lands End, August 2001

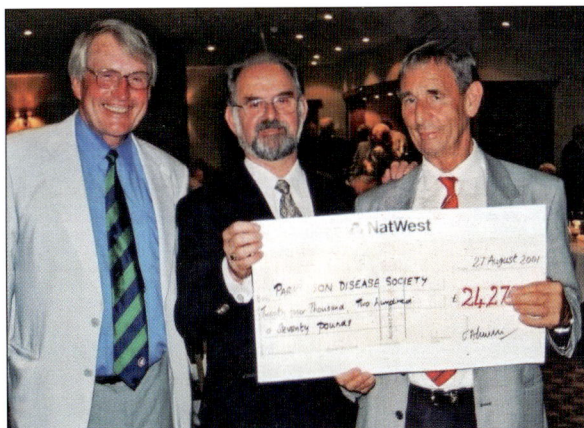

48. Ted Charlesworth presenting part of the money raised to
Ken Roache, local secretary to the Parkinson Society

49. Chester Golf Club - Captain's Day, 2004

50. With Barbara and Lady Captain, Sheila Watson, dressed up for the Annual Ball

51. Captain's Drive into office, 2004

52. The 50th anniversary of the first Annual Dinner of the Honourable Society of Newts inaugurated in 1956. Alnmouth Golf Club, Foxton Hall, Northumberland

53. Summer Newts Meeting at Hawkestone Park, 2004.
From the left: Tom Nicholson (Hon. Sec), Ian Shepherd, Ian Sutherland, Peta Sutherland, Robin Craven, Godfrey Clark, Ann Crossley, John Crossley, Dick Armstrong , John Cundall, Ian Rowlands, Robert Jackson, Maurice Donnelly, Duncan Robson, John Craven, Jill Robson

54. The old barn and cowshed before conversion, 2008

55. Spy Hill Lodge, 2011

56. Family group, 2010
Standing from the left: John, Robin, Peter, David, Barbara
Sitting: Valerie, Carolyn, Janet, Susan

57. From the left: Joe Bright, Brian McAdam and John Willis,
Pulmenach, Brittany 2002

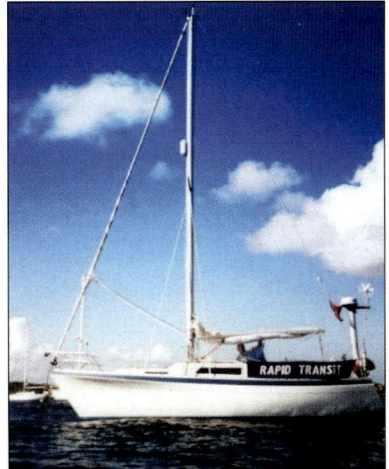

58. Rapid Transit, Scilly Isles 2002

59. Grandchildren 2009

From top left: Ben Craven,
Marc Sorrentino, Nicola Sorrentino,
Mike Craven, Caroline Sorrentino,
Zoe Sorrentino, Joanna Craven,
Charlie Craven, Ashleigh Craven,
Harry Craven, Jamie Craven

been brought up there where his father, a rather eccentric Italian, ran a garage and car hire business which he wanted Neville to manage. Reference has been made earlier to Sue's arrival in the world when we lived in Cheltenham. As the eldest and the only girl, (I write this on her forty-eighth birthday) she has a special place in our hearts, although we don't see as much of her as we would like now that she lives overseas. Her life has never been plain sailing. She is an extrovert (if sometimes occasionally OTT), has a loving nature and a determined will to ride out any storms which come her way. She is also exceptionally loyal. Sadly Sue's marriage failed a few years ago but she decided to stay in the BVI where she now works in a nursery with small children. She has four of her own upon whom she dotes. They all went to Rydal like their parents and the eldest two have recently graduated from Cardiff University. Marc has been doing journalism and hopes to transfer to Chester Law School. Nicola read law and criminology. Caroline is also at Cardiff and following her successful foundation year, hopes to study photography. Zoe has recently started a course on events management.

Peter married Carolyn Pott in 1992 who he also met while at Rydal. She is a qualified pharmacist and now works part time for the NHS. Bright, outgoing and highly organised she is the perfect match for Pete, our second child and eldest son who seems to have inherited the Cooper brains and the Craven brawn. He was Senior Prefect at school in his final year and decided to follow family tradition and study agriculture at Newcastle University. After graduating he joined Coopers Animal Health and for a while worked in South Wales on the sales side. He then moved to their head office in Crewe to specialise in marketing but left when they relocated to the south of the country. After a short spell in the computer industry and a redundancy looming, he decided to set up a marketing company with a business partner and they now employ about thirty staff with an office in the centre of Manchester. He is a natural leader, able communicator and possesses sufficient self-confidence to produce and act in local pantomime – the adult variety, at Chester Rugby Club. In last year's production he was a 'camp' tin man in *The Wizard of Oz*; previous to that he starred as *Jack and the Beanstalk's* decidedly odd mother. His friends eagerly await this year's event, whilst his wife and daughter shudder at the prospect. That the pantomime makes a significant contribution to club funds and junior tours is a welcome bonus. Pete had an impressive rugby career, captaining his university side and subsequently playing first division rugby at Orrel when the professional league system was introduced in the eighties. He captained the Chester club and also played for Cheshire on his return north. He now runs the under-sixteen junior section of the club within which Charlie, his elder son, is one of the prominent players. Pete and Carolyn have two other children: Jo, the eldest, is keen to become a doctor and is currently in the middle of her 'A' level course. She may even choose Newcastle University continuing a family tradition. Harry, at eleven the younger son, seems to have more than his fair share of talent and looks as if he will move effortlessly through life, especially on the sports field. The family have recently moved into Spy Hill farmhouse.

David was the first to get married at the tender age of just twenty. Like the others he had gone to Rydal as a boarder at thirteen. He found the academic work uninspiring and many of the school rules too restricting, although his enthusiasm for rugby was very evident. It became clear that staying on for the sixth form was not a realistic option. As our third child he was relatively quiet, cool-headed, determined and fiercely independent. His interest lay in practical farming and after a short spell at Reaseheath Agricultural College in Cheshire, he went to work for my old pal Dan Cherrington in Devon – a challenge if ever there was one. After marrying Valerie Knight, a teacher, he returned to Cheshire and started work at Grosvenor Farms, the Duke of Westminster's estate in Cheshire. He has been there ever since and has now become one of their key staff, specialising in building operations, although he can turn his hands to most things. They live in one of the big farmhouses with attached buildings where Val's various horses are stabled. She has an effervescent personality, is an accomplished eventer and passionate about animals in general and horses in particular. They have two sons, Ben and Mike, now in their twenties. Both have attended Hartbury College in Gloucestershire. Ben has an agricultural degree and has recently taken a job as an assistant farm manager in Lancashire, whereas Mike will graduate next year in business studies. He was awarded a sports scholarship and would prefer to negotiate a rugby contract with one of the premiership clubs rather than seek gainful employment at this stage. He was unlucky not to get selected for the England under-eighteen squad in 2009. Before returning to college in September 2010, Mike played for the Chester 1st XV in their opening three league matches, thrilling his grandfather and enabling him to boast to his friends that this was a club record as we were the only family to field a first team player over three successive generations. Dave played over 100 first team games and is now Head Coach to the Chester 1st XV.

Robin, our youngest son, had to fight for survival within a group of siblings only a few years older. A natural mimic with a great sense of humour and a tendency to exaggerate the good things in life, Rob can hold his own in any company. At Rydal, he struggled academically, suffering from undiagnosed dyslexia for some years. Determined and ambitious, he managed to get to Loughborough to study sports science, although he decided not to pursue that interest after finishing his course. Instead he went to New Zealand and soon began working for a company which manufactured artificial turf for sports surfaces. He returned to the UK to marry his girlfriend, Janet Blackburn, in 1994 and they then lived in Auckland for six years. The company decided to start a manufacturing site in the UK at Kidderminster and Rob accepted the challenge to help set up the factory and production line. He was then promoted to their research department and travelled extensively around the company locations worldwide. They have two bright children, Ashleigh, fourteen who was born in New Zealand and is into artistic things in a big way and eleven-year-old Jamie, our youngest grandchild. He is also a promising rugby player and aspiring golfer. Rob played a few games for Chester 1st XV before he left Cheshire and now helps coach juniors at his local club near Kidderminster. He's also potentially the best golfer in the family when he can manage to keep the ball on the course.

We enjoyed a number of extended family holidays as the grandchildren started arriving. The first was to Obergurgl in Austria over Christmas 1992 to celebrate our thirtieth wedding anniversary. We were a party of six small children and eleven adults, including Reg and Ann Knight, Val's parents. Rob was yet to marry. Skiing skills were variable from complete beginners to those who could cope fairly comfortably with red runs. The hotel was excellent, good snow, fantastic weather the whole week and much vigorous après ski. Reg was even encouraged to take over the microphone and give us an amazing rendering of 'My Way' which would have brought the tears to 'Ole Blue Eyes' himself.

In April 2001 we managed to get everyone together to visit Sue and Neville in Tortola in the British Virgin Islands. They lived with their four children in an impressive villa overlooking Long Bay on Beef Island. By then all the grandchildren had arrived, although Harry and Jamie were barely out of nappies. We travelled Air France and it was an exciting expedition getting them all through Charles de Gaulle airport and onto the flight across the Atlantic to the Caribbean. We swam, sailed, snorkelled, got sunburnt and drank 'bushwackers' as if there was no tomorrow. The sight of two naked small boys being thrown by their fathers from one side of the pool to the other, where they were caught and promptly returned by one of their uncles was one of the many memories of a fabulous couple of weeks.

We are indeed blessed with a large and loving family. We are immensely proud of them all and can look back on countless occasions at Spy Hill and other locations when we have been together to celebrate Christmas, a special event or simply to enjoy each other's company. But the mainstay in all of this has been Barbara. There is no one I know who leads such a full life. Her medical career still continues in that she provides locum sessions at the Hospice of the Good Shepherd near Chester. It is an inspiring establishment, largely funded through public donation and dedicated to palliative care for the terminally ill. She is Churchwarden at St John's, our local Anglican church in Ashton Hayes and is currently leading the search for a new vicar as the incumbent has recently retired. She is deeply involved in our local golf club, where she chairs the Social Committee and undertakes numerous other tasks on members' behalf. On a quiet moment she can be observed watering the overhanging baskets or weeding the flowerbeds. At home she runs the garden and co-opts me for the odd job – but she really prefers to do it herself. I mistake too many flowers for weeds. Above all she has been the guiding light in my life. When the big decisions have arisen, such as career promotions, the buying of the farm, the education of our children, the building of our new home – she has provided the vision, encouragement and commitment to go for it. Without that I truly believe I wouldn't have had the confidence to achieve many of the challenges related in this book. It is and has been a wonderful partnership.

This section would be incomplete without some mention of our dogs, past and present. I remember we had a golden Labrador called Sheena when we lived in Stanley almost seventy years ago and they have always been my favourite. Barbara and I both love dogs and decided that this breed had the right temperament to be entirely safe within a young family. Importantly, we also had the space to ensure a large dog could get plenty of exercise. Our present incumbent is Tilly who we were given by friend and former colleague John Butler. He had attempted to train her as a gun dog but she was hyperactive and it was timely that he was looking to find a good home for her. She has a propensity to chase lambs or eat hens, so we have to watch her carefully in the spring and barricade our flock of four hens into a dog proof compound – otherwise she is the typical soft, spoilt, hungry hound like the rest have been. Tilly's predecessor was Kiri and whilst her mother Poppy was alive, they reigned between them for about twenty years. There were three before that – Willoughby or Willy for short, named after a neighbouring admirer, Sally and Sandy. So a succession of six golden Labradors have given us much love, enjoyment, heartache and anxiety over the forty-five years we have lived at the farm. A black Labrador puppy called Ernie has recently taken up residence with the neighbours in the main house across the yard.

Losses

Sadly my brother Derek died in April 2004 aged seventy-three. Since he had divorced Ann, his first wife many years previously, I had seen less and less of him. He continued to live in Northumberland but seemed to want to change his lifestyle and find new friends after he married for the second time. It was a great pity as he had been such a popular figure in our rowing days. Like my father, he was a heavy pipe smoker and despite having had a heart bypass, continued the habit until his early seventies. But the damage had been done and he eventually died from heart failure. He had asked that his ashes be cast on the River Wear, the scene of some of his earlier rowing exploits. After his funeral, attended by Ann and scores of his old friends, Ian Shepherd and I rowed a boat down the river to Elvet Bridge and together with his eldest son Andrew who steered, deposited the contents of the urn into the dark waters. Derek and Ann had three children. Andrew, a banker, lives with his wife Julie and daughter in Essex. We had an enjoyable family reunion with them in 2009 when he celebrated his fiftieth birthday. We also recently attended another similar party in Yorkshire, where Tony lives with his wife Alison and family. Tony has a senior job in National Milk Records, so therein lies another link with the past. Linden, Derek's daughter has her own business making and selling jewellery up and down the land. She is an enterprising character and bought a 200-acre hill farm a couple of years ago. She and her partner Alex, a Spanish vet, live on the farm in the wilds of Northumberland and Ann, her mother, has a small cottage a couple of miles away.

Another major loss in our wider family has been Diana, one of Barbara's twin sisters. She had suffered from cancer for many years and died in 2007. I first met Di when she worked in the field laboratory at Cockle Park when I was doing my PhD in the early Sixties. She was a lovely person, but sadly her marriage to Martin Thompson, a Lincolnshire farmer, was also to fail. She had two sons, Michael who works in insurance and lives with his wife, Alison and their small children in Yorkshire. Philip the younger son now lives in New Zealand and looks set to remain there. His home in Nelson in the South Island is rapidly becoming a tourist centre for the travelling Cravens as they visit the country of their mother's birth. Phil and Emma have two small boys who are growing up as Kiwis – life goes in circles. Fiona, Di's daughter, is in personnel and lives in Lincoln with her partner Ian who is a teacher. Fiona is the life and soul of any party, tremendous company and always giggling.

Di's loss was a particular blow to Squint, her twin sister. Twins are said to have a special relationship with each other – it was very evident that this was the case between these two. Squint and Dai Morris continue to live in South

Wales where they retired after dairy farming replaced Dai's academic career at Aberystwyth. They have a small farm and a pedigree flock of sheep, well known in breeder and show circles. Their two children, Sally and David live nearby.

Reflections

The title of a book, *Letters to my Grandchildren*, written by Tony Benn, the Labour politician, attracted my attention as a possible format for writing this last chapter. I found the content disappointing but I am not one of his grandchildren – they may have an altogether different view. What I was hoping for was a guide as how best to express my thoughts about the various aspects of my life – lived forwards and understood backwards.

Whilst my children are now well into middle age, this book might resonate with their childhood and even fill in some background which is new to them. My grandchildren, however, range more widely in age from eleven to twenty-three and are just embarking on life's journey. They will soon, or already have, started formulating their own views of the world and the seemingly vast opportunities which appear within it. They are on the threshold of experience and have to think carefully about what they want to do and how they plan to live their lives. They have to decide about their values and the way they communicate with other people. As they grow older they will probably take more notice of their own age group than they will of parents and grandparents. We will appear out of touch and belonging to another era. Peer pressure is a potent influence and values have changed so much over the last few decades. Nevertheless, I hope all of these exciting young people find this chapter of some relevance in their thinking as we do have, whether we like it or not, a genetic link which must contribute in some way to their own make-up. These reflections are not supposed to be recommendations, although they may appear so, but simply opinions and conclusions based on my own experience.

I have chosen to divide the text into three broad categories. First, the basics which affect everyone. Second, the senses and some human emotions and third, the key factors which influence personality, whether genetic or environmental. It is a crude division and there will be overlap, but the method provides some structure to a complicated series of interlinked subjects.

Luck, life and death

Luck or as some may prefer to call fate, is my first basic category. Luck is independent of any action you might take to influence events as they directly affect you. The most extreme piece of luck might be the very fact of being born in the first place, considering the number of sperm involved in a single conception. Winning millions of pounds in a lottery doesn't qualify in this sense as at least you purchased a ticket. At the other extreme, being killed in an earthquake is pure chance, although one might argue that you can even have some influence over this in terms of where you choose to live. Timing is also a matter of luck in so far as it involves the individual. Someone born at the turn of the last century might have had no choice but to be on the front line of the Battle of the Somme in World War One – further bad luck would mean his remains lie in one of the war graves in France. It is our good luck that we have been born at a time in history when war and conscription have not disrupted our lives.

Other less extreme forms of luck, good or bad, occur every day in our lives. Some would say you make your own luck and to some extent this is true. You can improve the chances of something happening or not happening to you if you take certain lifestyle decisions, such as what you eat, whether to take exercise, whether to take drugs (nicotine and alcohol included) or the sort of life you lead. It certainly seems that some people are luckier than others, winning raffle prizes or sinking putts.

I don't suppose many people still accept the Old Testament book of Genesis as a literal explanation of how the world began, although no doubt there will be some extremists who will argue the case. It was not until the mid-nineteenth century with the publication of Charles Darwin's *Origin of the Species* that any scientific theory of life's origin began to be taken seriously. Now there is no doubt that this publication forms the foundation of evolutionary biology, but then the scientific establishment was closely tied to the Church of England. Ideas about transmutation between different forms of life were controversial as they conflicted with the belief that species were unchanging parts of a designed hierarchy and that humans were unique and unrelated to animals. As society becomes more secular and religion plays a diminishing role in people's lives, the concept of divine intervention as a cause of life becomes, for many, difficult to accept. Perhaps of more relevance is to ask the question – what, if anything, is the meaning of life? This has been the subject of philosophical study for centuries.

The explanation I like is summed up in *The Meaning of Life*. At the end of the film we can see Michael Palin being handed an envelope. He opens it and provides the viewers with 'the meaning of life':

Well it's nothing very special. Uh, try to be nice to people, avoid eating fat, read a good book every now and then, get some walking in and try to live in peace and harmony with people of all creeds and nations.

It's probably blasphemy in the eyes of the devout, but it seems to be a reasonable guideline to me.

Whatever the reasons or the meaning of life might be, we should live it to the full. Work and play hard; strive for success but not at the expense of other people; keep the importance of work in relation to other responsibilities in perspective and don't take life too seriously – keep a sense of humour.

Death is an altogether more defined condition and is often popularly described, along with taxes, as nothing more certain. Whilst every passing is a sad loss, real sorrow should be reserved for the few who die at a young age, perhaps with a family they don't see grow up, or who get killed in an accident.

Like the meaning of life, the real question about death is whether there is some form of life after it. The Christian religion requires the follower to believe in the life hereafter. No one can define what that might be even in a spiritual sense. It is part of Christ's teaching and expressed as faith – but that in my definition is a human emotion and one with which I have difficulty. I think it was Peter Ustinov who said: 'Children are the only form of immortality that we can be sure of.' I agree with that.

I have much sympathy for euthanasia where the quality of life has gone or the pain of living becomes unbearable. I don't doubt the legal dangers that exist, but as a society we should be able to move forward to a more realistic option for those who want to end their suffering.

Finally on this rather morbid subject, there is one responsibility that everyone should deal with once they have any worthwhile assets and that is to make a formal will. It's not painful, you won't die any younger, but in the event of unforeseen tragedy it will save much heartache for those you leave behind.

The senses

The vast majority of people are born with five senses, all in full working order, and no doubt there is an element of heredity in all of these. Sadly some are born without some of them or have impaired senses and start life at a severe disadvantage. The two main senses are sight and hearing. The others are touch, taste and smell. The way we develop the main senses and use them play a vitally important part in forming our personality and the impression we convey to other people. It is important to understand how they can be used to best advantage.

'Seeing is believing' is well known to be the only way many people are prepared to accept that something is real – other people's description can be exaggerated or misleading. To be observant is to use sight to its full potential, a vital component in the business of communication with other people. What people say is usually only a part of the opinion they really have. Body language is often more important than speech. Consider what you look for when you meet people for the first time. Do they look welcoming; do they look you in the eye; do they seem interested; do they proffer a handshake or a kiss and does it seem genuine rather than affected? How, therefore, do you think you appear to other people when they first meet you? Do you care? If you want to make an impression, then you need to think about what your body language tells them, not just what you say.

In the sporting sphere to have an 'eye for a ball' and to have peripheral vision are essential basics upon which to build. On their own they do not, of course, guarantee success in games or in anything else. Lots of other things are needed, such as balance, courage and anticipation to name but a few, but good sight is an essential start.

Hearing is very different from listening. Listening carefully to people (and watching body language) is the only way to understand what they are on about. A rule of thumb might be: you have one mouth and two ears – use them in that proportion. Just as important is their perception of you in terms of whether they think you are listening to what they say. Your own body language and reaction are important. Conversing without being able to observe, such as on the telephone is a poor, although essential, substitute for face-to-face conversation.

Whilst sight and hearing are the most important senses, touch, taste and smell make possible the enjoyment of many of the good things in life. The human touch; the feel of silk or Tilly's ears; the taste of new potatoes or a cold white sauvignon; the smell of the sea or the fragrance of a rose are just a few of my favourite things.

Emotions

The 'New Living Translation' of the final part of Corinthians 1, chapter 13 from the New Testament is probably one of the most widely read lessons in the modern Anglican Church:

> When I was a child, I spoke and thought and reasoned as a child. But when I grew up, I put away childish things. Now we see things imperfectly, like puzzling reflections in a mirror, but then we will see everything with perfect clarity. All that I know now is partial and incomplete, but then I will know everything

completely, just as God now knows me completely. Three things will last for ever – faith, hope and love – and the greatest of these is love.

In my definition, these words constitute three of the most important human emotions. I fully accept that the bible stories stem from the life of a remarkable man and over the centuries his teaching has been at the heart of Christianity, indeed the history of our civilised society derives most of its values from this source and I have tried to conform to those standards in my own life. But those with 'faith' truly believe that there exists some sort of spiritual life after death. I hope they are right but I remain unconvinced. Barbara, on the other hand, has faith and takes a very active role in our local church and village community. Her caring nature and medical experience maybe help her to carry this conviction through her own life. It is a subjective and personal question and one which only the individual can decide for him or herself.

Hope is said to 'spring eternal in the human breast'. 'Where there is life there is hope', is another oft quoted text, but sadly this is not always true. Seamus Heaney discusses the difference between hope and optimism:

> Optimism is a matter of personality; you are either optimistic or you are not. Hope, on the other hand, is something entirely different; something based on a foundation and on evidence of being hopeful or otherwise.

I support this view but fear I tend towards the pessimistic or downside of a situation, seeing the glass half empty rather than half full. Maybe it is a defence mechanism, psychologically preparing me for the worst outcome and initiating the planning necessary to limit damage and find a way back. People like Ted Charlesworth, of bike ride fame, had to bear the burden of so much ill health and yet did not despair or give up hope, providing an inspiration, certainly for me and for others to follow. However, bad things may seem hope only disappears when you decide something is hopeless. It very rarely is.

Love is perhaps the most difficult human emotion of all to explain and describe. My favourite reference on the subject is from Louis de Bernières' *Captain Corelli's Mandolin*. Dr Iannis counsels his daughter:

> And another thing. Love is a temporary madness, it erupts like volcanoes and then subsides. And when it subsides you have to make a decision. You have to work out whether your roots have so entwined together that it is inconceivable that you should ever part. Because this is what love is. Love is not breathlessness, it is not excitement, it is not the promulgation of promises of eternal passion,

it is not the desire to mate every second of the day, it is not lying awake at night imagining that he is kissing every cranny of your body. No, don't blush, I am telling you some truths. That is just being 'in love', which any fool can do. Love itself is what is left over when being 'in love' has burned away and this is both an art and a fortunate accident. Your mother and I had it, we had roots which grew towards each other underground, and when all the pretty blossom had fallen from our branches we found that we were one tree not two.

I am no romantic, as Barbara will readily testify. In fact I put in the category of luck that we ever met. Time and circumstance combine to bring people together. But whether luck or not, the outcome has been a love which has stood the test of time and for which I will be eternally grateful.

There are lots of emotions in life, both positive and negative. Fear is a major one and in its lower case is often experienced as worry or anxiety about what might happen rather than what is likely to happen. I do tend to worry too much about events or problems and have an inclination to bottle things up and deal with them on my own which has been unhelpful. It has to be better to talk to someone close and allow them to help you to get things into perspective. Things are never as bad as they seem.

Anger is another powerful negative emotion and it, too, sometimes leads on to a loss of temper when things can be said or done the individual may deeply regret once the storm is over. We need to keep a balance and remember how much our late friends would like to be able to do the things we now take for granted.

On the positive side, I rate empathy very highly. This means putting yourself in other people's shoes and thinking about how they feel about a situation. I once attended a seminar about motivation and the factors which influence success in business. Our group was asked to consider the single most important attribute needed as a manager to become successful. We came to no consensus having discussed leadership, delegation, sincerity, integrity and many others. Our lecturer advocated the skill of making others feel valued and important both within and outside an organisation. We all need a sense of achievement and encouragement in our daily lives and it is worth thinking about your own feelings when those in authority take time to say thank you or well done.

Personality – and work

Inherited traits such as size, features, eye, skin and hair colour are self-evident. Less obvious are congenital abnormalities, intelligence, mannerisms and coordinative skills important in sport and other physical activities. Through concentrated effort, individuals can improve their status in most of these facets but they need to work much harder than those blessed with natural talent. The fact that some who have this talent and yet choose not to exploit it is a waste of golden opportunity. Acquired personality traits evolve in individuals through learning and the experiences gained throughout life. These not only advertise to everyone else the sort of person you are but also need to be packaged skilfully when applying for a job. Clearly there is always a technical element to an interview and an examination of whether your qualifications meet the requirement, but whether you get a job offer also depends on the assessment of your personality.

Locating a job which meets your qualifications and appears to offer both the opportunity and the challenge to fulfil your ambitions is the first step on the road. Getting an interview comes next and then making the most of it completes the process, hopefully towards receiving an offer.

Completing the dreaded application form is an art in itself. There will, no doubt, be courses to help you complete these to best effect and this must be time well spent. A short covering letter no more than one page, maybe handwritten (if you have neat and legible writing) about why you want the job and how you hope to develop into it are worthwhile inclusions. This should not be a CV, such details are obviously important but can be included either on the application form or a supplementary sheet. Academic qualifications and practical experience may open the door but to get through it your CV needs also to include extramural activities such as sport, interests, overseas travel and voluntary work. If the job is worthwhile there are bound to be a lot of applicants and your application has in some way to stand out and attract attention. Captaining a team, being a school monitor or a secretary of a society, all show an aptitude towards taking responsibility.

Apart from a smart appearance, the interview itself is critical. Sometimes this is only half an hour or can extend to include overnight and involve all sorts of role playing designed to get the candidates to express their full potential. Whatever the format there are a number of key areas which the panel will be looking for.

Self-confidence is much valued. At the time you probably feel that this is the last thing you have in front of an interviewing panel. But if they are good at their job they will try and get you to relax and talk about yourself. Be yourself, don't try and pretend that you are different to what you are. Be bold and offer firm opinions based on your experience so far. Further into the discussion, they may want to see how you react to pressure. Try and keep calm and if you don't know the answer, say so – don't waffle.

Judgement is another skill which is highly valued. When confronted with a problem, what are the priority actions you need to identify? Can you see these for yourself or do you need to seek another opinion before deciding what action to take. There is no harm in sharing the problem with someone else, especially when you are unfamiliar with the subject, but choose the person carefully. It will be your judgement whether you take the advice or not.

Communication skills are essential for most jobs and therefore need to be at the forefront of your priorities throughout your education and training. If your accompanying letter has been produced skilfully, this will give the panel some guide as to how you write, but it is the way in which you project your personality and what you say during an interview which really matters. Remember to look the interviewers in the eye; be aware of what your body language tells them; be enthusiastic, honest and don't overdo it.

Speaking in public usually fills people with dread before they gain confidence. I certainly found it a tough challenge in my early career, and started by trying to ask intelligent questions at meetings, and followed it by accepting invitations to contribute or speak at farm discussion groups. It depends on the occasion, but if possible, it is much better to speak rather than read from a prepared text. This enables you to gauge audience reaction, an essential ingredient, if the delivery is to be effective. At first I tended to over-prepare with lots of overhead slides and notes to prompt my memory. These do provide some back up but the downside is if you have to refer to them you can rarely find your place. Better to have a list of one-liners or single words at hand (out of sight) to use in extremis. Speaking in public is a form of self-publicity, which will not appeal to everyone. It requires persistency and practice, but if you want to get on you have to become reasonably proficient.

One additional group of inter-related characteristics which I also regard highly in people is: trust, loyalty, sincerity and integrity. It is difficult to conceive of an individual having one of these without the others. In my view they are not only essential personality traits, they are also the values which matter most in real friendship.

Happiness is a state of mind rather than a personality trait, but however it is defined it is a goal most of us aspire to score. I offer three quotations which seem relevant to the subject and I sincerely hope you all succeed in finding it.

'Well', said Pooh, 'what I like best' and then he had to stop and think. Because although eating honey was a very good thing to do, there was a moment just before you began to eat it which was better than when you were, but he didn't know what it was called. (A.A. Milne).

If you observe a really happy man you will find him building a boat, writing a symphony, educating his son, growing double dahlias, or looking for dinosaur eggs in the Gobi desert. He will not be searching for happiness as if it were a collar button that has rolled under the radiator. He will not be striving for it as a goal in itself. He will become aware that he is happy in the course of living life 24 crowded hours of the day. (W. Bevan Wolfe)

Since you get more joy out of giving joy to others, you should put a great deal of thought into the happiness you are able to give. (Source unknown)

I would also recommend you read the poem 'If' by Rudyard Kipling. When I was a fourteen-year-old schoolboy I chose to read this during a recitation in front of the whole school. It was a terrifying experience and I didn't feature in the prize list. Perhaps not to everyone's taste, but it still seems to me, after all these years, to sum up many of the values we should strive for in life. The second verse reads:

If you can dream – and not make dreams your master;
If you can think – and not make thoughts your aim;
If you can meet with Triumph and Disaster
And treat those two imposters just the same;
If you can bear to hear the truth you've spoken
Twisted by knaves to make a trap for fools'
Or watch the things you gave your life to, broken,
And stoop and build 'em up with worn out tools.

The poem ends:

If you can fill the unforgiving minute
With sixty seconds worth of distance run,
Yours is the Earth and everything that's in it,
And – which is more – you'll be a Man, my son!

Finally during a trip to the BVI in January 2011, we visited an exclusive resort on Guana Island. During lunch a Chinese professor, an eminent horticulturalist, introduced himself. He was in charge of a small group who were developing a tropical garden on the island. A diminutive figure, he was all smiles of welcome and delighted that we had a real interest in his project.